Parametric Lie Group Actions on Global Generalised Solutions of Nonlinear PDEs

Mathematics and Its Applications

Managing Editor:

M. HAZEWINKEL
Centre for Mathematics and Computer Science, Amsterdam, The Netherlands

Volume 452

Parametric Lie Group Actions on Global Generalised Solutions of Nonlinear PDEs

Including a Solution to Hilbert's Fifth Problem

by

Elemér E. Rosinger

Department of Mathematics and Applied Mathematics,
University of Pretoria,
South Africa

Springer-Science+Business Media, B.V.

A C.I.P. Catalogue record for this book is available from the Library of Congress.

ISBN 978-90-481-5093-9 ISBN 978-94-015-9076-1
DOI 10.1007/978-94-015-9076-1

Printed on acid-free paper

The petal brave
A synchrony of light,
Angles, forms bewitch
The knight,
Onward joy
Revere a grail,
Non-linear-one
A-symmetry so sweet,
A beacon e'er
Shifting eye,
Geometrically composed !

Dedicated to
my wife Joselyn
with thanks for her
contribution

. . . The Whey so pure
In Numbers stilled
all anguish
Rilled,
Beyond the flare
A buffered muse !
Whose knightly tears
From Wandered-Stone
The Acorn-Willow
Breach is
Onward Rent
From painless burn
To seering sweet !

Table of Contents

Preface

This book presents a solution of the *harder* part of the problem of defining *globally* arbitrary Lie group actions on such *nonsmooth* entities as generalised functions. Earlier, in part 3 of Oberguggenberger & Rosinger, Lie group actions were defined globally - in the *projectable* case - on the nowhere dense differential algebras of generalised functions A_{Ω}, as well as on the Colombeau algebras of generalised functions, and also on the spaces obtained through the order completion of smooth functions, spaces which contain the solutions of arbitrary continuous nonlinear PDEs. Further details can be found in Rosinger & Rudolph, and Rosinger & Walus [1,2].

To the extent that arbitrary Lie group actions are now defined on such nonsmooth entities as generalised functions, this result can be seen as giving an answer to Hilbert's fifth problem, when this problem is interpreted in its original full generality, see for details chapter 11.

In this way, the present book is addressed to those who are interested in solving nonlinear PDEs, and in particular, in studying the Lie group symmetries of their classical or generalised solutions. Indeed, as it happens, quite basic nonlinear PDEs, such as for instance, the shock wave equation, have *nonprojectable* Lie group symmetries, as well as *generalised solutions* of physical interest, like the rarefaction waves and the Riemann solvers.

This book resulted from a more careful analysis of the nature of the difficulties when we want to go from the particular case of projectable Lie groups acting on algebras of generalised functions, to *arbitrary* Lie group actions on these algebras, and we want to define these actions *globally.*
And as a conclusion of this analysis, one is led to the *parametric* approach, both of Lie group actions, and of classical or generalised functions.

Rather surprisingly, this parametric approach proves, in our context, to have two important advantages, namely, one of *calculus*, and the other of *functorial* nature. The calculus advantage relates to the simple and well known fact that the partial derivatives of any order of a parametrically given function can be computed from it, without first having to bring the function to the usual, nonparametric form. The functorial advantage, relating perhaps even to a simpler fact, is the one which will actually allow the most easy, direct and natural global definition of arbitrary Lie group actions. In fact, as mentioned below, it will allow as well for the equally easy global definition of a far larger class of Lie semigroup actions.

As a general remark about the parametric approach to the global definition of arbitrary Lie group actions, it is rather ironic to note that, in an embryonic, partial and *local* manner, this approach has in fact been in use for a long time by now.

Indeed, given a smooth function $f : \Omega \longrightarrow \mathbf{R}$, with $\Omega \subseteq \mathbf{R}^n$ nonvoid, open, and given an arbitrary Lie group action $G \times M \longrightarrow M$, where $M = \Omega \times \mathbf{R}$, the usual way this Lie group action on M is extended to f is as follows. We consider the graph of f, that is, the set $\gamma_f = \{ \ (x, f(x)) \mid x \in \Omega \ \}$, which is obviously a subset of M. Therefore, for any $g \in G$, we can define pointwise the action $g\gamma_f$ and obtain again a subset of M.
Unfortunately, however, in general, it will *not* be true that

$$g\gamma_f = \gamma_h$$

for a certain smooth function $h : \Omega \longrightarrow \mathbf{R}$, which function h if it existed, it would obviously correspond to the *global* action of g on f, that is, we would have

$$gf = h$$

And then, the usual way in dealing with arbitrary Lie group actions on functions will overcome this difficulty simply by a *double* localization : g is taken near the identity $e \in G$, and in addition, f is restricted to suitable nonvoid, open subsets Δ of Ω.
It is clear, however, that the consideration of the graph γ_f of f amounts to replacing $f : \Omega \longrightarrow \mathbf{R}$ by the following special *parametric* form of it, namely $f_* : \Omega \longrightarrow M$, where $\Omega \ni x \longmapsto f_*(x) = (x, f(x)) \in M$. Furthermore, in this case $g\gamma_f$ is nothing else but gf_*, which can *always* be defined globally, irrespective of the function f, or of the Lie group action G on M.
And thus it becomes clear that the *only* difficulty we have ever faced with defining globally arbitrary Lie group actions on functions is *not* at all related to Lie groups or functions, but solely to our rather unformulated, and yet quite implacable *intent* to have gf_* retranslated into a usual, *nonparametric* function $h : \Omega \longrightarrow \mathbf{R}$.

On the other hand, in this book, the parametric approach to Lie group actions is adopted and pursued in its *full* extent, this being the simple and fundamental reason for the fact that *arbitrary* Lie group actions can now be defined *globally* on classical and generalised functions.

Finally, as a rather unexpected *bonus*, the functorial nature of the parametric approach to Lie group actions will in fact allow as well the definition of arbitrary smooth - thus typically *noninvertible* - actions. Such noninvertible actions can, of course, no longer belong to Lie group actions, but only to Lie *semigroup* actions. The interest in such Lie semigroups of actions comes from the fact that they range over a significantly *larger* class of actions. Therefore, when applied to the study of solutions of PDEs - this time as *semisymmetries* - they can offer new additional insights. Furthermore, as pointed out by P J Olver, semigroups of actions appear quite naturally in several aspects of the classical Lie theory. See for details chapter 13.

And now, to the organization of the book.
Chapters 1 - 5 motivate, introduce and present basic applications of the parametric approach to global Lie group actions on smooth functions. This forms the stepping

stone towards the full development in chpaters 7, 9 of the parametric approach to global Lie group actions on generalised functions.

In this book, we consider generalised functions from the so called nowhere dense differential algebras A_Ω. These were the first differential algebras of generalised functions in the literature, and were introduced in the 1970s, see Rosinger [3-10], Oberguggenberger [1], Colombeau. The reasons we consider the generalised functions in the nowhere dense differential algebras A_Ω are detailed in chapters 1, 12. They are mainly related to the fact that, among all other later algebras of generalised functions, the algebras A_Ω enjoy a rather simple interplay between ring theoretic and topological type properties, *without* however requiring any kind of growth type conditions, which in applications would inevitably bring with them significant difficulties. Moreover, these algebras A_Ω are the only ones known so far, which have a *flabby* sheaf structure. And this property is fundamental in order to be able to accommodate the largest class of *singularities* so far, which the generalised functions can have. Namely, these singularities can be on arbitrary closed and nowhere dense subsets. And two of their features are particularly important : first, such subsets can have arbitrary *large* positive Lebesgue measure, and second, in their neighbourhood, the generalised functions are *not* restricted in any way. A result of this is the global version of the Cauchy-Kovalevskaia theorem, which so far could only be obtained in the algebras A_Ω, see subsection 7.5.2.

Chapter 7 starts, step by step, the extension of the parametric approach to arbitrary and global Lie group actions on generalised functions in the algebras A_Ω. And soon it becomes obvious that, in order to obtain a natural enough construction, one has to go beyond the usual, nonparametric form of the algebras A_Ω, and use still more large spaces, this time of *parametric* generalised functions which need no longer form algebras. In this respect, the obvious complexity of the various constructions in chapter 7 can be seen, among others, as a further - and particularly detailed - motivation of the fully fledged and unrestricted parametric approach, which is then presented in chapter 9. This complexity is specially evident related to the operator δ in (7.4.46), which is *only* needed if we still insist that the results of arbitrary Lie group actions on generalised functions should be retranslated into the usual, *nonparametric* form. And to further highlight the questionable nature of such an insistence, we should note that, in general, it will only lead - and it can only lead - to *local* results. This fact is precisely the reason for the particularly involved nature of the mentioned operator δ, which does nothing else but to collect together *all* of the local results of the nonparametric approach.
The various constructions in chapter 7 make use of the results in the Appendix, where the issue of the local invertibility of local diffeomorphisms is studied in detail.

After these complications in chapter 7, the free use of the parametric approach is finally presented in chapter 9, where very general results on the Lie symmetries of generalised solutions of large classes of nonlinear PDEs are given.
It should, however, be noted, recognized and admitted that there exists a certain rather well entrenched and widespread *reluctance* in dealing with parametrically given functions. Therefore, the mentioned unformulated and implacable tendency

towards dealing with usual, *nonparametric* functions, or at least, towards ending up with results given by such functions.
And yet, the results on the parametric approach presented in chapters 5, 9, if considered more carefully, can to a good extent release us from that reluctance.

Chapters 6, 8, and section 3 in chapter 9, apply the parametric approach to *non-projectable* Lie group symmetries of the rarefaction waves and Riemann solvers of the nonlinear shock wave equation. Chapter 6 is of a preparatory nature, chapter 8 illustrates some of the complexities of the approach in chapter 7, if trying not to depart to far from the usual nonparametric approach, while finally, in section 3 of chapter 9, the free application of the parametric approach is presented in its simplicity and power.

With chapters 1 - 9 and the Appendix, the main body of the book is completed. The rest of the book, that is, chapters 10 - 13, are dedicated to related issues.

Chapters 10, 11 present the connection with Hilbert's fifth problem.
Chapter 12 describes in some detail the unique position of the nowhere dense differential algebras of generalised functions A_Ω among all other known differential algebras of generalised functions.

Finally, chapter 13, as a mere 'introduction to a future introduction', tries to motivate the researchers in Lie groups to pursue what to the author appears to be a surprisingly fortunate, and also quite significant opening, namely, the theory, or rather, the future theory of *genuine* Lie semigroups of actions. And to try to further emphasise the point, the author would like to state here that he would not be surprised, if this future theory of genuine Lie semigroups of actions proved to be quite a bit more important than the rest of the work in the present book.

As a conclusion, one may see both the parametric approach to Lie group actions on classical and generalised functions, as well as the future theory of genuine Lie semigroups of actions as a more *globally* minded approach to large classes of *singularities* of classical smooth, or nonsmooth generalised functions.

At last, a word about LAT_EX.
After many years in which I published mathematics papers and books, it was for the first time with this book that I typed it myself, thanks to having taken the gap of learning in the process just about enough from the wonders of LAT_EX.
At first, LAT_EX appears to be quite forbidding indeed, reminding those among us who have been around for longer of that ... super hyper programming language of the exuberant late 1960s, called PL1, a language meant once and for all to ... end all programming languages, and which just like other similar outrageously complex and involved contraptions of a not yet well understood and digested technology, was bound to collapse under its own clumsy burden, before it could ever really take off ...
However, as soon as one manages to muster the typing in LAT_EX of just a bit of mathematics, one quite clearly awakes to the feeling that, just as with the xerox

copier, the fax, the e-mail and the www, it is really hard to understand how could we ever live without the *LATEX* ...

E E R

Pretoria, April 1998

Chapter 1

Introduction

The Main Results. A novel approach to Lie group actions on usual and generalised functions, based on *parametric representation*, is introduced. This allows, for the first time, an easy *global* definition of arbitrary nonlinear, including *non-projectable*, Lie group actions on functions, as well as on generalised functions, and in particular, on the traditional Schwartz distributions.
In short, parametrisation means globalization in the study of Lie group actions on usual and generalised functions.

The extension of arbitrary Lie group actions to large classes of generalised functions also means that an answer is in this way obtained to Hilbert's fifth problem, when this problem is considered in its full generality. This answer extends earlier similar ones given in the particular case of projectable Lie group actions, see Oberguggenberger & Rosinger, Rosinger [11], Rosinger & Rudolph, and Rosinger & Walus [1,2].
It is also shown that the usual Lie group symmetries of classical solutions of $\mathcal{C}^\infty$-smooth nonlinear partial differential equations will as well remain Lie group symmetries of the generalised solutions of such equations.

Applications to the nonprojectable Lie group symmetries of the rarefaction waves and of the Riemann solvers of the nonlinear shock wave equation are presented.

As a rather unexpected bonus of the parametric approach introduced in this work, it will be possible to start a theory of *genuine* Lie semigroup actions, that is, of Lie semigroup actions which are *not* subsemigroups of Lie group actions. The interest in such a theory is in the *vast* extension it can offer in the analysis of symmetries, or rather *semisymmetries* of functions, generalised functions, and thus, of solutions of linear and nonlinear partial differential equations.

Further, as will be seen, the parametric approach allows an easier study of contact transformations.

The main results on the new, parametric approach to arbitrary Lie group actions on classical and generalised solutions of nonlinear PDEs are presented in Theorem 1, in chapter 5, and in Theorems 2, 3, and Corollaries 1, 2, in chapter 9. All these results show that the actions of the usual Lie group symmetries of nonlinear PDEs upon their solutions will again give solutions. In this regard, the result in Theorem 1, in chapter 5, shows that property in the case of classical solutions given in a parametric form. The other mentioned four results in chapter 9 show the same for generalised solutions which are in parametric form.

Parametrisation : Two Advantages. The use of parametric representation of usual and generalised functions in the study of Lie group actions, introduced in this work, proves to have two important advantages, both of them facilitating a *global* approach, an approach which, since Chevalley, is considered to be the modern one.
The first advantage is of functorial or categorial nature, and it offers the possibility of a most simple, natural, general, unrestricted and global definition of arbitrary nonlinear Lie group actions on usual functions, see (4.1) - (4.4). In fact, it allows for the easy definition of the vastly more large set of *semigroup* actions on usual and generalised functions. This latter case of actions, however, will be the object of a subsequent work, see comments at the end of chapter 4, as well as an introductory presentation in chpater 13.
The second advantage is of a simple, calculus nature and it is about the ease in the global computation of the derivatives of any order of a parametrically given function. This proves to be of a particular interest in the study of contact transformations as well, see comments at the end of chapter 5.

Functorial Advantage. The functorial advantage comes from two facts. First, in a parametrically given function there is no distinction made between independent and dependent variables. Second, for the same given function, one can introduce the new variables, that is, the parameters, in a large variety of ways.
The idea to dispense with our usual attachement to the distinction between independent and dependent variables was suggested in a different context, see Olver [1, p. 222], where quotient manifolds were defined, in order to provide a rigorous formulation of the basic method for finding group invariant classical solutions for partial differential equations. That approach does not involve parametrisation, and it does not make use of the functorial advantage.
In this work, however, the functorial advantage of parametric representation is used in order to define, and then employ in several nontrivial applications, arbitrary nonlinear Lie group actions on usual and generalised functions.

Calculus Advantage. There exists a well known, elementary, yet rather neglected property of functions given by a parametric representation. Indeed, the derivatives of any order of such a parametrically given function can easily, *explicitely* and *globally* be computed, without first having to eliminate the parameters, and thus revert to the usual definition of a function, that is, to revert to a mapping from its domain of independent variables, to its range of dependent variables. Furthermore, as is known, the class of parametrically given functions is wider than that of usual ones.

On the other hand, the elimination of parameters, needed when we want to obtain the usual form of the function, is often rather difficult, and it can only be done locally.

Parametrisations. In its essence, the parametric method amounts to the following *embedding* of the usual definition of a function into a larger concept. We note that a usual function

$$A \ni x \stackrel{f}{\longmapsto} y = f(x) \in B$$

is actually *constrained* to be a correspondence from the set A of its independent variable x, to the set B of its dependent variable y.
On the other hand, a parametric representation of f can be given by any *pull back* type mapping

$$P \ni p \stackrel{h}{\longmapsto} h(p) = (x(p), y(p)) \in A \times B$$

which maps any suitably given parameter domain P into the graph of f, under the following two conditions : $y(p) = f(x(p))$, for $p \in P$, and the mapping $P \ni p \longmapsto x(p) \in A$ is surjective.
With respect to P, this, in general, only implies that its cardinal is not smaller than that of A.
However, in this work, further suitable topological conditions will be required on the parameter domain P, while the parametrisation h will be assumed $\mathcal{C}^\infty$-smooth.

It follows that, in general, a parametric representation will introduce an *additional* variable p, ranging over P, which this time will be mapped into the pair $(x(p), y(p))$ of the original independent and dependent variables, pair which is an element in the cartesian product $A \times B$.

This kind of embedding, obtained by introducing an additional variable, and thus going beyond the constraint of only dealing with the usual independent and dependent variables, proves to have an important and naturally built in advantage. Namely, it allows for the first time - and in a straightforward manner - the *global* definition of arbitrary nonlinear, including nonprojectable, Lie group actions on functions, see (4.1) - (4.4). Indeed, this approach is able to take into account the fact that, in the study of partial differential equations, the relevant Lie groups do in general act *jointly* on the pair of independent and dependent variables, that is, on elements of the cartesian product $A \times B$, rahter than on each of them separately, see (1.2).
In the usual nonparametric approach, however, when we want to define the Lie group action on a function, and obtain again a function, we cannot do so, unless at the end we are able to *separate* the independent and dependent variables, by expressing the latter as a function of the former. And in the nonprojectable case of Lie group actions, this will often not be possible, except locally in the independent variable, and also, near to the trivial, identical Lie group transformation.

On the other hand, if we start, and end, with parametrically given functions, then as seen in (4.3), (4.4), we shall have no difficulties at all.
And when in addition we make use of the mentioned, and rather neglected, property of parametric representation with respect to explicit computation of derivatives, we can obtain in a natural manner the results in this work.

First details with respect to the way we use in this work the advantage of the possibility of explicit computation of derivatives of parametric functions are presented in chapter 5.
Here we should recall that the above mentioned property of parametrically given functions, when it comes to the explicit computation of their derivatives, also happen with implicit and inverse functions, and it is the result of the *chain rule*. And as is known, the chain rule is but the nonlocal consequence of its local, pointwise version, which follows directly from the linear algebra rule for *matrix multiplication*, Rudin.

Classical Lie Group Actions. In the case of nonlinear PDEs of the general form

$$(1.1) \quad T(x,D)U(x) = 0, \quad x \in \Omega \subseteq \mathbf{R}^n$$

where Ω is nonvoid open, $U : \Omega \longrightarrow \mathbf{R}$ is the unknown function and $T(x,D)$ is a $\mathcal{C}^\infty$-smooth nonlinear operator, the relevant Lie groups G act on the manifold $M = \Omega \times \mathbf{R}$, according to

$$(1.2) \quad \begin{array}{l} G \times M \longrightarrow M \\ (g,(x,u)) \longmapsto g(x,u) = (g_1(x,u), g_2(x,u)) \end{array}$$

where $x \in \Omega, u \in \mathbf{R}$ are the independent and dependent variables, respectively, and

$$(1.3) \quad \begin{array}{l} G \times M \ni (g,(x,u)) \longmapsto g_1(x,u) \in \Omega \\ G \times M \ni (g,(x,u)) \longmapsto g_2(x,u) \in \mathbf{R} \end{array}$$

with g_1 and g_2 being $\mathcal{C}^\infty$-smooth.

It will be assumed that the nonlinear partial differential operator in (1.1) is of the general form

$$(1.4) \quad T(x,D)U(x) = F(x, U(x), \ldots, D_x^p U(x), \ldots), \quad |p| \leq m$$

where $m \in \mathbf{N}$, and F which is $\mathcal{C}^\infty$-smooth, are arbitrary given.

We note that, given $g \in G$, in view of the Lie group axioms, it follows that the mapping

$$(1.5) \quad M \ni (x,u) \stackrel{g}{\longmapsto} g(x,u) \in M$$

is a $\mathcal{C}^\infty$-smooth diffeomorphism.

Lie Group Actions on Generalised Functions. A main aim of this work is to extend the *arbitrary* nonlinear Lie group actions (1.2) to global generalised solutions U on Ω of the nonlinear PDEs in (1.1). This extension will then contain the respective results in Oberguggenberger & Rosinger, Rosinger [11], Rosinger & Rudolph, and Rosinger & Walus [1,2], where nonlinear Lie group actions, in the *projectable* case, see (2.3) below, were defined on the mentioned generalised solutions. A short account of these earlier results is presented in chapter 10.

We recall that the mentioned global generalised solutions are elements of *differential* algebras of generalised functions constructed in Rosinger [1-10]. A class of such differential algebras used in this paper, and called the *nowhere dense algebras*, is given by

$$(1.6) \quad A = \mathcal{A}/\mathcal{I}_{nd}(\Omega)$$

where $\mathcal{A}$ are suitable subalgebras in $(\mathcal{C}^\infty(\Omega))^{\mathbf{N}}$ and $\mathcal{I}_{nd}(\Omega)$ is a specially chosen *nowhere dense ideal* in $(\mathcal{C}^\infty(\Omega))^{\mathbf{N}}$. Here we denoted by $\mathcal{C}^\infty(\Omega) = \mathcal{C}^\infty(\Omega, \mathbf{R})$ the set of all $\mathcal{C}^\infty$-smooth functions $U : \Omega \longrightarrow \mathbf{R}$, while $(\mathcal{C}^\infty(\Omega))^{\mathbf{N}}$ is the set of all sequences of such functions. Clearly, just like $\mathcal{C}^\infty(\Omega)$, so is $(\mathcal{C}^\infty(\Omega))^{\mathbf{N}}$ an associative, commutative and unital algebra, when considered with the usual termwise operations on sequences of functions. In this way, the nowhere dense differential algebras of generalised functions (1.6) are but suitable *quotient algebras of subalgebras* in $(\mathcal{C}^\infty(\Omega))^{\mathbf{N}}$.
Furthermore, as seen below in (1.7), the nowhere dense algebras (1.6) contain the Schwartz distributions.

To be more precise, in the monograph and two papers mentioned, see also chapter 10, projectable nonlinear Lie group actions were defined on the differential algebras (1.6). Furthermore, it was shown that the usual Lie group symmetries of classical solutions of nonlinear PDEs remain Lie group symmetries of the generalised solutions of these equations, provided that the respective Lie groups were projectable. The construction in the present work extends these results to arbitrary, that is, possibly *nonprojectable*, nonlinear Lie group actions on the same algebras of generalised functions.

Answer to Hilbert's Fifth Problem. One effect of this extension is that it gives an answer to Hilbert's fifth problem, when this problem is considered in the full generality of its original formulation, namely, related to Lie group actions on manifolds, and not only on the respective Lie groups themselves. This answer to Hilbert's fifth problem, presented in chapter 11, comes to strengthen the one given earlier in the monograph and two papers mentioned, where only the case of projectable nonlinear Lie group actions was dealt with.

Why Use the Nowhere Dense Differential Algebras of Generalised Functions ? The nowhere dense differential algebras A in (1.6), independently of possible Lie group actions on them, are of interest owing to at least two reasons. First, they satisfy a requirement which in the PDE community happens to be held up as a sort of sine qua non, namely, they contain the Schwartz distributions, that is, we have the inclusion

(1.7) $\mathcal{D}'(\Omega) \subseteq A = \mathcal{A}/\mathcal{I}_{nd}(\Omega)$

and this inclusion preserves both the usual multiplication and differentiation of $\mathcal{C}^\infty$-smooth functions. Moreover, these differential algebras allow within themselves for a very large range of nonlinear operations to be performed on the Schwartz distributions and other similar generalised functions. In particular, any nonlinear operation defined by $\mathcal{C}^\infty$-smooth functions, or even by certain classes of generalised functions, can be performed within these differential algebras.

Second, they are the *only* class of differential algebras of generalised functions known so far, such that, as a kind of roof theory, they can treat in a conveniently unrestricted and *unified* way both the regular and singular phenomena of interest related to the solutions of such large classes of nonlinear PDEs as in (1.1), (1.4). And the reason for that is in the way these algebras are defined, *without* any kind of growth conditions involved, and in terms only of the usual topology on $\mathbf{N} \times \Omega$, and of the ring structure of $\mathcal{C}^0(\mathbf{N} \times \Omega)$.

Third, the nowhere dense algebras form a *flabby* sheaf, unlike the Colombeau, or many other algebras of generalised functions, or for that matter, the Schwartz distributions. And this flabbiness proves to have important applications, see Malios & Rosinger, or Kaneko.

These nowhere dense differential algebras of generalised functions were first introduced and used in Rosinger [3-6], and their full theory is presented in Rosinger [6-8].

It should be mentioned that in the late 1970s, an awareness emerged about the need to incorporate large enough classes of *singularities* into new spaces which appear in the algebraic-differential structures underlying an intrinsic approach physics, see Souriau [1,2], and for more recent inputs, Heller [1,2], and also Kirillov [1,2]. Indeed, the traditional spaces of generalised functions, such as for instance, the Sobolev spaces, or the Schwartz distributions, in addition to their disadvantage of typically being merely linear spaces and not algebras, could only accommodate a limited class of singularities. In particular, the singularites had to be concentrated on sets of Lebesque measure zero, and near them, one had to impose various limiting growth conditions on usual and generalised functions.
The mentioned interest in the suitable incorporation of large classes of singularities arose in particular in the more abstract approaches to differential geometry, aimed to be applied, among others, in de Rham cohomology , which is an essential componenet of an intrinsic treatment of physics, see von Westenholz, or in geometric quantization, see Mallios [5], Brylinski, Blattner, Kostant, or for a more basic text, Simms & Woodhouse.

From this point of view, the nowhere dense differential algebras of generalised functions in (1.6) prove to have the capability to deal, so to say, in a 'smooth' manner with such a rather large class of singularities, as those concentrated on arbitrary *closed, nowhere dense* subsets $\Gamma \subseteq \Omega$.
In fact, owing to the structure of the *nowhere dense ideals* $\mathcal{I}_{nd}(\Omega)$, the algebras (1.6) simply do *not* notice, are *not* sensitive to, or shall we say, *jump* over all singularities on closed, nowhere dense subsets of their domain of definition.

And it should be noted that such a property in handling singularities is nontrivial, since closed, nowhere dense subsets can have *arbitrary* large positive Lebesgue measure, Oxtoby.
Clearly, the case of such singularities on subsets Γ of positive Lebesgue measure cannot be treated by the Schwartz distributions, and in particular, by Sobolev spaces. Equally, they cannot be treated by the generalised functions in the Colombeau algebras. See further details in chapter 12, and also in Grosser et.al.

One of the immediate consequences of this 'smooth' treatment of any closed and nowhere dense singularity is the development in Mallios & Rosinger, which allows a far reaching extension of the de Rham cohomology, with the incorporation of singularities situated on the mentioned kind of arbitrary closed and nowhere dense subsets in Euclidean spaces.
An earlier consequence of this 'smooth' method was the *global* Cauchy-Kovalevskaia theorem, see Theorem G C-K in subsection 7.5.2, established in Rosinger [7-9]. According to that theorem, arbitrary nonlinear analytic PDEs have global generalised solutions given by elements of the algebras (1.6). Furthermore, these global generalised solutions are usual analytic functions on the whole of Ω, except for closed nowhere dense subsets, which in addition, and if desired, can be chosen so as to have zero Lebesgue measure.
Such a type of very general nonlinear and global existence result has not been obtained in the Colombeau algebras of generalised functions, owing among others to the special *growth conditions* which appear essentially in the definition of those algebras, see for instance Colombeau.

It can also be mentioned that the differential algebras (1.6) were the first in the literature to have the traditionally required property (1.7), and together with other similar large classes of algebras, they were introduced starting back in the 1960s, see Rosinger [1-8], and the respective reviews in MR 92d:46098, Zbl. Math. 717 35001, MR 92d:46097, Bull. AMS vol.20, no.1, Jan 1989, 96-101, and MR 89g:35001.
Furthermore, in Rosinger [6], a wide ranging and purely algebraic, ring theoretical characterisation was given for the first time for all possible differential algebras of generalised functions which, as in (1.7), contain the Schwartz distributions. Also, a number of properties of various classes of such algebras were presented.
The Colombeau algebras, Colombeau, by necessity were shown to be a particular case of such classes, see Rosinger [7,8].
Finally, the result on the existence of global generalised solutions for arbitrary nonlinear analytic PDEs, used in subsection 7.5.2, could so far be obtained only within the nowehre dense algebras (1.6), Rosinger [7-9].

In connection with the nowhere dense differential algebras A in (1.6), let us mention two more facts relevant to this work.

First, it is *nontrivial* to prove that, as seen in (1.7), these differential algebras contain the Schwartz distributions. Details in this respect can be found in Rosinger [1-10]. Comments on the desirability, or otherwise, of such embeddings of distributions are presented at the end of this chapter, as well as in chapter 12.

On the other hand, it proves to be surprisingly *easy* to define arbitrary nonlinear Lie group actions on these differential algebras, provided that the parametric approach is used. In the particular case of projectable Lie group actions, the usual nonparametric approach as well allows an even more direct and easy definition of such actions on the nowhere dense differential algebras of generalised functions (1.6), see Oberguggenberger & Rosinger, and for a short account, chapter 10.

However, the fact that *nonprojectable* Lie group actions, as defined in this work on generalised functions, can be applied to the rarefaction waves and the Riemann solvers of the nonlinear shock wave equation is again a *nontrivial* result.

Open Problems. As mentioned, a main aim of this work is to extend the results in Oberguggenberger & Rosinger, Rosinger [11], Rosinger & Rudolph, and Rosinger & Walus [1,2] to arbitrary nonlinear, including nonprojectable, Lie group actions on generalised functions.
Once that is accomplished, as in this work, then the next stage, which at present remains one of the important open problems, is to try to find *nonclassical* Lie group symmetries for generalised solutions of nonlinear PDEs. For that purpose, one will have to solve the following open problems, namely, to define, and then integrate, *vector fields of generalised functions*. A start in this direction has already been made in Hermann & Oberguggenberger, and Kunzinger.

Needless to say, in the difficult venture of solving in a type independent manner large classes of nonlinear PDEs, the method of *relating* solutions to transformations which lead again to solutions is of particular importance and usefulness.
However, so far, in line with Lie's theory, the transformations considered have been limited to *invertible* ones, that is, to transformations which can be incorporated into *groups*.
On the other hand, the functorial advantage resulting from the way Lie group actions are defined in (4.1) - (4.4) on parametric functions, allows in exactly the very same way, and without the need for any additional complication, the definition of actions which no longer correspond to transformation groups, but to the *vastly* larger class of transformation *semigroups*. Therefore, this extension gives us what appears to be a significant opportunity in the further and deeper study of solutions of large classes of nonlinear PDEs. A start towards this study is presented in chapter 13.
Finally, the ease in the *explicit* computation of derivatives of any order of parametrically given functions gives the opportunity to deal with contact transformations of such functions in a more convenient way, see the end of chapter 5.

And here, if we are speaking about open problems, then perhaps, we could also mention one which needs no particular elaboration about its long outstanding and fundamental importance, and one which, to some extent, may hopefully be helped by the developments in the present work :

Towards a General Relativistic Quantum Theory ? In his 1972 paper entitled 'Missed Opportunities', see pp. 645-647, Dyson draws attention to the following state of affairs, and his comment is still relevant today :

" . . . The most glaring incompatibility of concepts in contemporary physics is that between Einstein's principle of general coordinate invariance and all the modern schemes for quantum mechanical description of nature. Einstein based his theory of general relativity on the principle that God did not attach any preferred labels to the points of space-time. This principle requires that the laws of physics should be invariant under the Einstein group E, which consists of all one-to-one and twice differentiable transformations of coordinates. By making full use of the invariance under E, Einstein was able to reformulate the whole of classical physics (electromagnetism and hydrodynamics) in E-invariant fashion, and so determine unambiguosly the mutual interactions of matter, radiation and gravitation within the classical domain. There is no part of physics more coherent mathematically and more satisfying aestetically that this classical theory of Einstein based upon E-invariance.
On the other hand, all the currently viable formalisms for describing nature quantum-mechanically use a much smaller invariance group ...
In practice all serious quantum-mechanical theories are based either on the Poincare group P or the Galilei group G. This means that a class of preferred inertial coordinate-systems is postulated a priori, in flat contradiction to Einstein's principle. The contradiction is particularly uncomfortable, because Einstein's principle of general coordinate invariance has such an attractive quality of absoluteness. A physicist's intuition tells him that, if Einstein's principle is valid at all, it ought to be valid for the whole of physics, quantum-mechanical as well as classical. If the principle were not universally valid, it is difficult to understand why Einstein achieved such deeply coherent insights into nature by assuming it to be so . . . "

Certainly, the theory of Lie group actions, be it on manifolds, or on classical or generalised functions, is one of the basic mathematical theories which deal with a variety of concepts of invariance.
As is known, most of the laws of physics happen to be expressed as equations, which quite often are nonlinear partial differential equations. And many of the relevant solutions of such equations, because of internal mathematical, or perhaps other reasons, exhibit singularities. Therefore, they cannot be dealt with by classical, sufficiently smooth functions. However, the invariance principles, such as for instance, that of Einstein's coordinate invariance in general relativity, cannot be seen as only applying to classical, smooth solutions.
In this way, there exists a significant interest in extending such invariance to large classes of nonclassical, nonsmooth, or in short, *singular* solutions. And so far, by far the *largest* such class of singularities dealt with in the literature - namely, on arbitrary closed and nowhere dense subsets, which therefore, can have arbitrary large positive Lebesgue measures - are those of the generalised functions in the nowhere dense differential algebras. These algebras are in the present work the new domain, upon which the actions of arbitrary nonlinear, including nonprojectable Lie groups are now extended.

Coming back to general relativity, recently, a number of researchers have started to use certain classes of generalised functions, in order to model singularities which

appear in connection with the Riemannian metric, or in the solutions of Einstein's field equations.
In this regard, one should perhaps not miss mentioning first the 1987 paper of Geroch & Traschen, which showed what goes *wrong*, if we try to use *linear* distribution theory in general relativity.
At the other end, we should mention the soon to appear Kluwer monograph by Kunzinger, et.al., which in its Part IV presents a state of the art account of the application of recent results in nonlinear theories of generalised functions, and in particular, of the Colombeau algebras, to general relativity.
Fortunately, the recent literature on such subjects shows a certain increase both in quantity an depth of enquiry. Here we shall mention some of the more relevant among those which the present author happens to be familiar with : Steinbauer [1,2], Kunzinger & Steinabuer, Clarke & Vickers & Wilson, Wilson, Clarke [1-3], Balasin [1-3], Aichelburg & Balasin [1,3], Balasin & Nachbagauer.
Finally, one should also mention the far deeper, and general approach in Mallios [6], which applies the latest results in abstract differential geometry, originated and developed in Mallios [1-5], see also Mallios & Rosinger. The approach to abstract differential geometry in Mallios [1-5] is accomplished *without any notion of differentiability*, as it identifies, and then uses the underlying essential algebraic structures involved, such as the de Rham complex, for instance. Therefore, for the first time in any known version of differential geometry, it offers a vast capability to incorporate *singularities*. And such a major advantage in the incorporation of singularities is indeed achieved, based on sheaf cohomolgy, and it starts by replacing the usual smooth functions by sheaves of algebras on arbitrary paracompact spaces. This is then, in short, the approach which allows for large classes of singularities to be easily incorporated in a differential type calculus which can reproduce many of the fundamental features of usual differential geometry, features of fundamental importance in general relativity, for instance, such as *connections* and *curvature*, see also recent related work in Vassiliou [1-4].

In view of these recent contributions, we are looking forward to the time when the 'missed opportunity' mentioned in the above citation from Dyson may be taken up.
And no doubt, at such a time, owing to the *essentially nonlinear* nature of general relativity, one can expect that the bringing together into an appropriate conjunction of singularities and invariance will no longer be possible to be missed ...

Acknowledgements. Grateful thanks are due to M Oberguggenberger for his appreciation of the ideas and results in this work. The same grateful thanks go to M Kunzinger for his careful reading of parts of various versions of the manuscript of this work, and for his useful criticism, comments and suggestions.

The author is particularly indebted to J P Olver for his private correspondence related to certain natural ways in which Lie semigroups of actions can appear.

My good old friend A Mallios happened again to visit me at the Univerity of Pretoria during the writing of this work. On that occasion, he brought to my attention a whole range of ideas and results, some recent, other ones less so, related to the issue of suitably general methods for the incorporation of *singularities* into any differential type theory which, among others, may aim to help in solving large classes on nonlinear PDEs. Here, I am particularly glad to express my gratitude for the respective discussions.

Greatful thanks are due to R Hermann, who expressed his appreciation of the ideas in this work, and in particular, suggested their possible extension towards contact transformations.

Part of the work, in particular, on Lie semigroups, was completed while the author was, for two months towards the end of 1997, visiting at the Erwin Schroedinger International Institute for Mathematical Physics of the University of Vienna, Austria, at the workshop on generalized functions. The author is indebted to M Oberguggenberger, W Thirring and P W Michor for the respective invitation, and to the latter, for some revealing discussions.

On the occasion of that workshop, M Hazewinkel, on a shorter visit to the mentioned institute, has - to the great delight of the author - expressed his interest in the present work. The author owes him grateful thanks.

By the way, one of the remarkable moments of the workshop happened during an informal talk of M Kunzinger on various attempted ways to extend the construction of certain algebras of generalised functions, and in particular, of the Schwartz distributions, to manifolds. Suddenly, and so surprisingly, M Hazewinkel asked the question :

Why should one consider the Schwartz distributions, when constructing such extensions to manifolds ?

Indeed, for most of us in the PDE venture, and in particular, in algebras of generalised functions, who have grown up in the era of that very strong paradigm of the Sobolev spaces, and later, of the Schwartz distributions, such a question would appear as purely ... unthinkable ...
And yet, there we were, with the question asked in front of us, and asked in a serious and competent manner ...
And all we could do was to throw that more than six decades old, very strong paradigm at it, in order to dismiss it out of hand as being a mere strange question, or, who knows, on the contrary, to try really to have a rather radically fresh look at the things, and ponder that question quite carefully, and also, for quite longer ...

Well, I myself chose the second option, an option which, in fact, was not so new to me, having thought about it earlier, on more than one occasion. What was nevertheless so freshly new even for me, was the sudden and quite blinding force of the conviction with which M Hazewinkel made all of us face that question in a public situation.
And indeed, all of the differential algebras of generalised functions ever constructed since the 1960s, had one common overall aim. Namely, to be able - unlike most of Sobolev spaces, and unlike the Schwartz distributions - to set up a proper *nonlinear* framework for a general enough study of nonclassical solutions of nonlinear PDEs.

And yet, just as in (1.7) above, we kept imposing on our own new nonlinear thinking some sort of particular, and not at all well thought out, type of ... coherence with the old linear spaces of generalised functions ...

Now of course, some such sort of coherence may be quite useful. However, what the question of M Hazewinkel tried so clearly and loudly to tell us was that, when going from the rather cosy world of linear theories, to the incomparable, and so far, mostly uncharted complexities of nonlinear theories, one should better avoid any hasty prejudice. In particular, one should refrain from reducing any possible and yet not known forms of coherence, to a mere inclusion, as in (1.7) for instance, and then impose it universally ...

Indeed, a quite well known saying some years ago had it that, there was plenty room in Hilbert spaces ...

But now, we should better stop for a moment, and realize quite deepley that, there is going to be plenty complexities, and therefore, surprises in nonlinear theories ...

Chapter 2

Actions on Functions, Difficulties

A first basic problem in Lie group theory, when applied to PDEs, is how to *extend* the action in (1.2) of the Lie group G on the manifold M, to an action of G on the $\mathcal{C}^\infty$-smooth functions

$$(2.1) \quad U : \Omega \longrightarrow \mathbf{R}$$

or more generally, on $\mathcal{C}^\infty$-smooth functions

$$(2.2) \quad U : \Delta \longrightarrow \mathbf{R}$$

where $\Delta \subseteq \Omega$ is nonvoid, open. And unless one solves this problem, one cannot speak about the Lie group invariance of classical, let alone, generalised solutions of PDEs.

From this point of view, the Lie group actions (1.2) are divided in two types, Olver [1,2]. The simpler ones, called *projectable*, or *fibre preserving*, satisfy the condition, see (1.3)

$$(2.3) \quad g_1(x,u) = g_1(x), \quad g \in G, (x,u) \in M$$

The special interest in Lie group actions (2.3) comes from the fact that they allow an easy *global* extension to action on $\mathcal{C}^\infty$-smooth functions.
Indeed, in this case, in view of (1.5), it follows that for $g \in G$, we obtain the $\mathcal{C}^\infty$-smooth diffeomorphism

$$(2.4) \quad \Omega \ni x \xmapsto{g_1} g_1(x) \in \Omega$$

Now, given $g \in G$ and U in (2.2), it is easy to define the Lie group action

$$(2.5) \quad g\,U = \bar{U} : \bar{\Delta} = g_1(\Delta) \longrightarrow \mathbf{R}$$

by

(2.6) $\quad \bar{U}(g_1(x)) = g_2(x, U(x)), \; x \in \Delta$

Indeed, (2.4) implies that in (2.5), we have $\bar{\Delta} \subseteq \Omega$ nonvoid, open, while (2.6) is equivalent with

(2.7) $\quad \bar{U}(\bar{x}) = g_2(g_1^{-1}(\bar{x}), U(g_1^{-1}(\bar{x}))), \; \bar{x} \in \bar{\Delta}$

However, an arbitrary Lie group action (1.2) need *not* be projectable. And in such a case the global extension of the Lie group action (1.2) to $\mathcal{C}^\infty$-smooth functions (2.1), (2.2) will typically *fail*. In this way, we are obliged, Olver [1,2], to limit ourselves to *local* Lie group actions on functions, and thus return to the pre Chevalley stage of Lie group theory.

Indeed, in the case of general, nonprojectable Lie group actions (1.2), we may immediately run into the problem of the possible *noninvertibility* as a $\mathcal{C}^\infty$-smooth function of certain $\mathcal{C}^\infty$-smooth mappings involved in the definition of the group action $g\,U = \bar{U} \; : \; \bar{\Delta} \longrightarrow \mathbf{R}$.

Let us illustrate this phenomenon in more detail. Given $g \in G$, let us write (1.3) in the form

(2.8) $\quad \begin{aligned} \bar{x} &= g_1(x, u) \\ \bar{u} &= g_2(x, u) \end{aligned}$

where $(x, u), (\bar{x}, \bar{u}) \in M$. Given $U \; : \; \Delta \longrightarrow \mathbf{R}$ as in (2.2), the natural way to define the group action $gU = \bar{U} \; : \; \bar{\Delta} \longrightarrow \mathbf{R}$ would be by the relation, see (2.8)

(2.9) $\quad \bar{U}\,(g_1(x, U(x))) = g_2(x, U(x)) \; , \; x \in \Delta$

which means that $\bar{U}(\bar{x}) = \bar{u}$. However, in order that (2.9) be a correct definition, we have to be able to obtain $x \in \Delta$ as a $\mathcal{C}^\infty$-smooth function of $\bar{x} \in \bar{\Delta}$, by using the first equation in

(2.10) $\quad \begin{aligned} \bar{x} &= g_1(x, U(x)) \\ \bar{u} &= g_2(x, U(x)) \end{aligned}$

and thus by replacing $x \in \Delta$ in the second equation above, in order to obtain $\bar{u}$ as a function of $\bar{x}$, that is, the relation (2.9). Furthermore, one also has to obtain $\bar{\Delta} \subseteq \Omega$ as being nonvoid, open. The crucial issue here is, therefore, the *$\mathcal{C}^\infty$-smooth invertibility* of the mapping

(2.11) $\quad \Delta \ni x \stackrel{\alpha}{\longmapsto} g_1(x, U(x)) \in \Omega$

which obviously depends on g and U. And as seen in the example next, this in general is not possible.

Example 1

Let us consider the following *nonprojectable* case of the Lie group action (1.2), where

$\Omega = \mathbf{R}, M = \Omega \times \mathbf{R} = \mathbf{R}^2, G = (\mathbf{R}, +)$ and for $\epsilon = g \in G = \mathbf{R}$, $(x, u) \in M$, we have

$$\bar{x} = x + \epsilon u^2$$
$$\bar{u} = u$$

Let us take $\Delta = \Omega = \mathbf{R}$ and the simple function $U : \Delta \longrightarrow \mathbf{R}$ defined by $U(x) = x$, with $x \in \Delta$. Then (2.11) becomes

$$\mathbf{R} \ni x \stackrel{\alpha}{\longmapsto} x + \epsilon x^2 \in \mathbf{R}$$

which is *not* invertible as a function, let alone as a $\mathcal{C}^\infty$-smooth function, except for the trivial group action corresponding to $\epsilon = 0$, that is, to the identical group transformation.

□

The usual way to deal with this situation, Olver [1,2], see also a recent development in Rosinger [12], is to consider the group action (1.2), as well as the mapping α in (2.11), and therefore $U : \Delta \longrightarrow \mathbf{R}$, *only locally*, that is, to restrict all of them to such suitable neighbourhoods of the neutral element $e \in G$, as well as of points $x \in \Delta$, on which α is $\mathcal{C}^\infty$-smooth invertible.

It is useful to note however that, depending also on the function U in (2.2), the mapping α in (2.11) can sometime happen to have a global, and not only local $\mathcal{C}^\infty$-smooth inverse, even in the case of a nonprojectable Lie group action. For instance, this happens if in the above Example 1, we consider $\bar{x} = x + \epsilon u$.
Let us here mention for further reference what happens in such a case, when the mapping α in (2.11) is invertible, regardless of the Lie group action being projectable or not, and when nevertheless its inverse α^{-1} is also a $\mathcal{C}^\infty$-smooth mapping. Then we can indeed turn to (2.9) in order to define the group action $g\ U = \bar{U}$ by

(2.12) $\quad \bar{U}\ (\bar{x}) = g_2(\alpha^{-1}(\bar{x}), U(\alpha^{-1}(\bar{x})))$, with $\bar{x} \in \bar{\Delta}$

where

(2.13) $\quad \bar{\Delta} = \alpha(\Delta)$ is open

Obviously, the case of *projectable* Lie group actions in (2.3) - (2.7) is included in (2.12), (2.13).

As mentioned in chapter 1, the introduction, in this work we shall take a *new route*, when dealing with the difficulties in (2.8) - (2.11), which we face in the case of general, *nonprojectable* nonlinear Lie group actions (1.2). This new route will *not* require the above mentioned traditional localisation of $g \in G$, α or U. In other words, we shall be able to perform *globally* arbitrary nonlinear Lie group actions on functions U considered defined on the whole of their unrestricted, original domains. Fortunately, this construction will quite easily extend to arbitrary nonlinear Lie group actions on generalised functions in the nowhere dense algebras.

Furthermore, this new route will be particularly simple and applicable without any undue restrictions.

A Simple, Basic Observation. The basis upon which we can delvelop this global approach is the following rather simple observation :
The traditional impediment which prevents us from extending arbitrary Lie group actions (1.2) to global actions on functions (2.2) is *not* related to Lie groups, but to our usual way of representing functions, by discriminating between independent and dependent variables. Once we do away with such a discrimination, by using a parametric representation of functions, the way to a natural and easy global Lie group action on functions is open.
Furthermore, as mentioned, the way becomes open to the far larger class of global Lie semigroup actions on functions, as well as to an easier approach to contact transformations.

Chapter 3

Parametric Representation of Functions

Need for Global Approach. It is useful to give another example, which by its most familiar setup, can further highlight the rather basic and simple, yet extreme difficulties we may face when trying to define *globally* the action of a *nonprojectable* Lie group on a function.

Example 2

We consider the Lie group action given by the *rotation of the plane.* In terms of (1.2), it means that $\Omega = \mathbf{R}, M = \Omega \times \mathbf{R} = \mathbf{R}^2, G = (\mathbf{R}, +)$ and for $\theta = g \in G = \mathbf{R}, (x, u) \in M$, we have

$$\begin{aligned} \bar{x} &= x\cos\theta &- \ u\sin\theta \\ \bar{u} &= x\sin\theta &+ \ u\cos\theta \end{aligned}$$

therefore, here again, we are dealing with a nonprojectable Lie group action. Let $\Delta = \Omega = \mathbf{R}$, and $U \ : \ \Delta \longrightarrow \mathbf{R}$ be given by the *parabola* $U(x) = x^2$, with $x \in \Delta$. Then (2.11) takes the form

$$\mathbf{R} \ni x \stackrel{\alpha}{\longmapsto} x\cos\theta - x^2\sin\theta \in \mathbf{R}$$

which, again, is *not* invertible, except for the trivial group actions, for which $\theta = k\pi$, with $k \in \mathbf{Z}$.

□

It follows that, except for a rotation of $\theta = \pm\pi$, which in this case amounts to a symmetry with respect to the origin of coordinates, the parabola $\Delta = \mathbf{R} \ni x \longmapsto x^2 \in \mathbf{R}$, when taken as a whole, *cannot* be rotated at all in the plane, without ceasing to be the *graph* of a function from $\Delta = \mathbf{R}$ to $\mathbf{R}$. Yet it is clear that, as a geometric object, by arbitrarily rotating in the plane a parabola, we again get a parabola.

Therefore, the difficulty must lie with the particular way we happen to *represent* the parabola, that is, as a function from $\Delta = \mathbf{R}$ to $\mathbf{R}$.

Parametric Representation. It turns out that the alternative way to represent functions $U : \Delta \longrightarrow \mathbf{R}$ in (2.2), namely, *parametrically*, will help us avoid the above difficulties related to the possible *lack of $\mathcal{C}^\infty$-smooth inverse* of the mapping in (2.11).
What we mean by parametric representation of a function in (2.2) is presented now. Let us recall here that parametric representation is a rather familiar method in differential geometry, among others, where it is used to define, for instance, the concept of submanifold, Kahn.

Given a $\mathcal{C}^\infty$-smooth function

$$(3.1) \quad U : \Delta \longrightarrow \mathbf{R}$$

where $\Delta \subseteq \mathbf{R}^n$ is nonvoid, open, we denote its *graph* by

$$(3.2) \quad \gamma_U = \{(x, U(x)) | x \in \Delta\} \subseteq M$$

A *parametric* representation of U is given by *any* $\mathcal{C}^\infty$-smooth function

$$(3.3) \quad \Lambda \xrightarrow{V} M$$

where $\Lambda \subseteq \mathbf{R}^n$ is nonvoid, open, such that

$$(3.4) \quad V(\Lambda) = \gamma_U$$

As seen in (3.22) - (3.25) and (3.31) - (3.34), later in this chapter, the above condition (3.4), although seemingly quite weak, has certain useful implications.

It is important to note that in (3.3), the set Λ of parameters is assumed to be n-dimensional. This however, is in line with the fact that the domain of definition Δ of U in (3.1) is also n-dimensional. In particular, since M in (3.2) is n+1-dimensional, condition (3.4) is quite natural. Later, when in (3.6), we define the class of arbitrary parametrically given functions, which are of interest in this work, we shall hold to this assumption on the dimension of the set of parameters.

Canonical Parametrisation. Clearly, an immediate, simple and natural parametric representation of U in (3.1) is given by

$$(3.5) \quad \begin{array}{ccc} \Delta & \xrightarrow{U_*} & M \\ x & \longmapsto & (x, U(x)) \end{array}$$

and we shall call U_* the *canonical* parametric representation of U. However, it is obvious that a function U in (3.1) can have many other parametric representations (3.3), (3.4). Details in this respect can be found at the end of this chapter. In particular, we shall see in (3.22) and (3.24) that in a certain sense U_* is the *simplest possible* parametric representation of U.

Classes of Parametrisations. Clearly, the set of functions in (3.3) is *larger* that that in (3.1). More precisely, not every function V in (3.3) is the parametric representation of a function U in (3.1). For instance, a nontrivially rotated parabola can be written as a function in (3.3), but not as a function in (3.1).

Let us, therefore, denote by

$$(3.6) \quad \mathcal{C}_n^\infty(M)$$

the set of all $\mathcal{C}^\infty$-smooth functions $V : \Lambda \longrightarrow M$, where $\Lambda \subseteq \mathbf{R}^n$ is nonvoid, open, and call them n-dimensional *parametric representations* in M. Also, let us denote by

$$(3.7) \quad \mathcal{C}_{par}^\infty(\Omega, \mathbf{R}) = \mathcal{C}_{par}^\infty(\Omega)$$

the set of $\mathcal{C}^\infty$-smooth *partial* functions $U : \Delta \longrightarrow \mathbf{R}$, see (3.1), where $\Delta \subseteq \Omega$ is nonvoid, open. Then (3.5) yields an *embedding*

$$(3.8) \quad \mathcal{C}_{par}^\infty(\Omega) \ni U \longmapsto U_* \in \mathcal{C}_n^\infty(M)$$

while on the other hand, in view of (3.3), (3,4), we obtain a mapping

$$(3.9) \quad \mathcal{C}_{par}^\infty(\Omega) \ni U \longmapsto \mathcal{P}_U \subseteq \mathcal{C}_n^\infty(M)$$

where we denoted by $\mathcal{P}_U$ the set of mappings V in (3.3), which satisfy (3.4). In other words, $\mathcal{P}_U$ is the set of all parametric representations of U. And in view of (3.5), it is clear that

$$(3.10) \quad U_* \in \mathcal{P}_U \ , \text{ with } U \in \mathcal{C}_{par}^\infty(\Omega, \mathbf{R})$$

hence $\mathcal{P}_U \neq \phi$.

The *important* point for our purposes is that the construction of *arbitrary nonlinear* Lie group actions on the set of functions $\mathcal{C}_n^\infty(M)$ will *no longer* suffer from the above difficulties related to the possible *lack* of a *$\mathcal{C}^\infty$-smooth inverse* of the mappings in (2.11).

Similar to (3.7), it will be useful, for $\ell \in \mathbf{N}$, $\ell \geq 1$ and $N \subseteq \mathbf{R}^\ell$ nonvoid, open, to denote by

$$(3.11) \quad \mathcal{C}_{par}^\infty(\Omega, N)$$

the set of all partial functions $U : \Delta \longrightarrow N$ which are $\mathcal{C}^\infty$-smooth, where $\Delta \subseteq \Omega$ is any nonvoid, open subset. Obviously $\mathcal{C}_{par}^\infty(\Omega, M) \subseteq \mathcal{C}_n^\infty(M)$.

Comparing Parametrisations. Here we further clarify the meaning of the parametric representation of functions, defined in (3.1) - (3.6).
Let us define a *preorder* $\leq$ on $\mathcal{C}_n^\infty(M)$, that is, a reflexive and transitive binary relation, as follows. Given $\Lambda \xrightarrow{V} M$ and $\Lambda' \xrightarrow{V'} M$, with $\Lambda, \Lambda' \subseteq \mathbf{R}^n$ nonvoid, open, then

(3.12) $\quad V \leq V'$

if and only if there exists a surjective $\mathcal{C}^\infty$-smooth mapping $\Lambda \xrightarrow{\varphi} \Lambda'$ such that the diagram is commutative

(3.13)

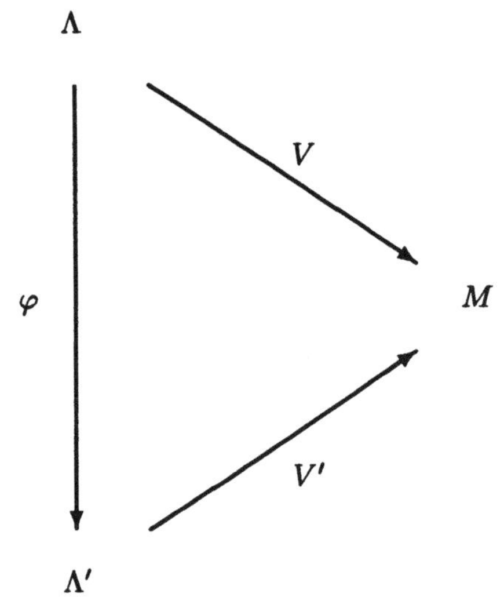

It is easy to see that, owing to the surjectivity of φ, we obtain

(3.14) $\quad V(\Lambda) = V'(\Lambda')$

A natural interpretation of this preorder $V \leq V'$ is that the parametrisation V' is *simpler* that V. This is illustrated in

Example 3

Let $\Lambda = \Omega \subseteq \mathbf{R}^n$ be nonvoid, open. Then let us take any $\varphi \;:\; \Omega \longrightarrow \Omega$ which is $\mathcal{C}^\infty$-smooth, surjective, but it is not injective. Also, let us take any $\mathcal{C}^\infty$-smooth $U \;:\; \Omega \longrightarrow \mathbf{R}$. We can now define the parametric function in $\mathcal{C}^\infty_n(M)$, namely

(3.15) $\quad \Lambda = \Omega \ni x \overset{V}{\longmapsto} V(x) = (\varphi(x), U(\varphi(x))) \in M$

Then clearly

(3.16) $\quad V \leq U_*$

However

(3.17) $\quad U_* \not\leq V$

indeed, assume $\Omega \xrightarrow{\Psi} \Lambda$ is surjective $\mathcal{C}^\infty$-smooth, such that $U_* = V \circ \Psi$. Then

$$x = \varphi(\Psi(x)), x \in \Omega$$

which means that, contrary to the assumption, φ is injective, since Ψ is surjective. □

Recalling that, see (3.5)

$$(3.18) \quad \Omega \ni x \overset{U_*}{\longmapsto} (x, U(x)) \in M$$

and comparing it with (3.15), where φ can be arbitrary under the mentioned assumptions, it follows that a natural meaning of (3.16), (3.17) is that the canonical parametric representation U_* of U is *simpler* than *all* the other parametric representations of U, given by V in (3.15), see also (3.22) and (3.24) below.

Basic Properties. It is useful to note that the simple looking condition (3.4) is precisely the one which leads to the situation in Example 3. Indeed, let $U : \Delta \longrightarrow \mathbf{R}$ be a $\mathcal{C}^\infty$-smooth function, with $\Delta \subseteq \Omega$ nonvoid, open, and let $\Lambda \overset{V}{\longrightarrow} M$ be any function in $\mathcal{C}^\infty_n(M)$, which therefore acts according to

$$(3.19) \quad \begin{array}{l} \Lambda \ni y \longmapsto V(y) = (V_1(y), V_2(y)) \in M \\ \Lambda \ni y \longmapsto V_1(y) \in \Omega \\ \Lambda \ni y \longmapsto V_2(y) \in \mathbf{R} \end{array}$$

Then it is easy to see that the following four conditions are equivalent:

(3.20) V is a parametric representation of U

$$(3.21) \quad V(\Lambda) = \gamma_U$$

$$(3.22) \quad V \leq U_*$$

(3.23) V_1 is surjective and $V_2 = U \circ V_1$

In view of the above and (3.9), it follows that

$$(3.24) \quad \begin{array}{l} \forall \ V \in \mathcal{P}_U : \\ \quad V \leq U_* \end{array}$$

In this way, in view of (3.22) and (3.24), and in the sense of Example 3, it is clear that for any given function U, its *canonical* parametric representation U_* is *simpler* than any other parametric representation of that function. Moreover, given two $\mathcal{C}^\infty$-smooth functions $U_1, U_2 : \Delta \longrightarrow \mathbf{R}$ in $\mathcal{C}^\infty_{par}(\Omega)$, then

$$(3.25) \quad U_{1*} \leq U_{2*} \Longrightarrow U_1 = U_2$$

which shows to what a large extent the canonical parametric representation does in fact determine a function.

Staying with Usual Functions. We conclude that given $\Lambda \overset{V}{\longrightarrow} M$ in $\mathcal{C}^\infty_n(M)$, then a sufficient condition for the existence of a function $U : \Delta \longrightarrow \mathbf{R}$ in $\mathcal{C}^\infty_{par}(\Omega)$, such that V is a parametric representation of U, is given by, see (3.23)

(3.26) $\Lambda \xrightarrow{V_1} \Delta$ is a $\mathcal{C}^\infty$-smooth diffeomorphism

In this case it also follows that, see (3.19)

(3.27) $U = V_2 \circ V_1^{-1}$

as well as, see (3.22)

(3.28) $V \leq U_*$

However, when in the subsequent chapters we deal with arbitrary nonlinear, and possibly nonprojectable Lie group actions on functions, we shall encounter the general situation of mappings $\Lambda \xrightarrow{V} M$ in $\mathcal{C}_n^\infty(M)$ which may fail to satisfy (3.26). This condition (3.26) can be seen as the more general reformulation of the $\mathcal{C}^\infty$-smooth invertibility problem in (2.11). The general case of parametric representation, including that of the generalised functions in the nowhere dense differential algebras (1.6), when condition (3.26) can easily fail, will be dealt with in subsection 7.4.

The main idea in this regard is *simple*, and as seen in (A.20), (A.30) and (A.39) in the Appendix, it is rather *inevitable*. Namely, whenever (3.26) fails, we shall instead look for *all* nonvoid open subsets $\Lambda' \subseteq \Lambda$ and $\Delta' \subseteq \Delta$, such that

(3.29) $\Lambda' \xrightarrow{V_1} \Delta'$ is a $\mathcal{C}^\infty$-smooth diffeomorphism

and then, correspondingly, construct

(3.30) $U = V_2 \circ V_1^{-1} \ : \ \Delta' \longrightarrow \mathbf{R}$

Equivalently, and in order not to carry *redundancies* which are obtainable simply by the *restriction* of the respective functions to nonvoid open subsets of their respective domains, we shall upon convenience also consider (3.29) and (3.30) for those nonvoid open subsets $\Lambda' \subseteq \Lambda$ only, which at the same time are *maximal*, see (A.6), (A.30) and (A.39) in the Appendix.

These constructions will then be extended to equivalence classes of sequences of parametric representations, and in particular, to the elements of the nowhere dense differential algebras (1.6).

Equivalent Parametrisations. Let us consider two parametric functions $V : \Lambda \longrightarrow M$ and $V' : \Lambda' \longrightarrow M$ in $\mathcal{C}_n^\infty(M)$, see (3.6). We say that V and V' are equivalent, and write

(3.31) $V \approx V'$

if and only if, see (3.4)

(3.32) $V(\Lambda) = V'(\Lambda')$

Clearly, if we have a usual function U in (3.1), then in view of (3.4), V will be a parametric representation of U, if and only if, see (3.5)

(3.33) $\quad V \approx U_*$

Also, owing to (3.12) - (3.14) and (3.32), it follows that

(3.34) $\quad V \leq V' \Longrightarrow V \approx V'$

Parametric Functions and a Two Track Approach. Throughout this work, we shall pursue a *two track* approach, when dealing with arbitrary nonlinear, including nonprojectable, Lie group actions on functions and on generalised functions.

Namely, for the sake of following the tradition, we shall first present the usual, *nonparametric* approach, and do so as far as one can pursue it, before one may be overtaken by technical difficulties which are no longer in balance with the results they provide. These kind of difficulties obtain a certain illustration in sections 7.2 - 7.5, as well as in chapter 8.

Then, as the second track, we shall go over to the *parametric* approach. Here we should mention that, contrary to certain negative expectations, one in fact loses very little, while on the other hand, one gains a significant ease, and also, the maximum generality. A good illustration about the power and advantage of this second track, parametric approach one can see in the results in Theorem 1, in chapter 5, and in Theorems 2, 3 and Corrolaries 1, 2, in chapter 9.

Let us mention here an illustration in the case of usual functions represented parametrically. First, in (4.1) - (4.4), arbitrary Lie group - and in fact, Lie semigroup - actions are defined. Then in (5.38), the easy and explicit way for computing globally the derivatives of parametric functions is presented. Further, (5.42), (5.43) give the definition for such parametric functions to be solutions of PDEs in (1.1), (1.4). Also, in (5.51) - (5.53), and in full detail, in Theorem 1, at the end of chapter 5, one can see the way arbitrary nonlinear Lie group actions transform parametric solutions into parametric solutions.

The usual, first, that is, nonparametric approach can indeed lead to considerable technical complications. However, it still deserves a certain attention, and not only due to tradition, but also to the fact that it allows for a discourse which is confined to the usual independent and dependent variables alone.
On the other hand, in view of its functorial, natural, simple, easy and very general character, see (4.1) - (4.4), we shall also present the way arbitrary nonlinear Lie groups act on parametrically given usual and generalised functions.
However, the main reason for the presentation of this second, that is, parametric approach is in the results it leads to, results which supplement both by power and simplicity those obtained through the first, usual, nonparametric approach. In particular, we shall be able to see in chapter 9 the power and simplicity of the parametric method, when applied to the nonprojectable Lie group symmetries of the nonlinear shock wave equation, as they act on rarefaction waves and Riemann

solvers. This power and simplicity is the more evident, when we compare these results with those in chapter 8, where the usual, nonparametric approach was employed.

Chapter 4

Actions on Parametric Representations

Natural Definition. Now with the use of parametric representations, we come to the *basic idea* of this work, namely we can define the arbitrary nonlinear Lie group actions on functions

(4.1) $\quad G \times \mathcal{C}_n^\infty(M) \longrightarrow \mathcal{C}_n^\infty(M)$

in the following simple and natural way. Given $g \in G$ and a function $\Lambda \xrightarrow{V} M$ from $\mathcal{C}_n^\infty(M)$, we define

(4.2) $\quad g\, V = g \circ V$

where in the right hand term, g is the mapping in (1.5). In other words, we use as definition of the Lie group action the commutative diagram

(4.3)
$$\begin{array}{c} \Lambda \xrightarrow{V} M \xrightarrow{g} M \\ \underbrace{\qquad\qquad\qquad}_{gV} \end{array}$$

Clearly, with the definition (4.2), (4.3), we have

(4.4) $\quad gV \in \mathcal{C}_n^\infty(M), \quad \Lambda \xrightarrow{gV} M$

that is, gV and V have the *same* domain of definition Λ, and the *same* range M.

Properties. We show now that the Lie group actions (4.1) contain as a particular case the usual Lie group actions on functions, Olver [1,2]

$$G \times \mathcal{C}^{\infty}_{par}(\Omega) \longrightarrow \mathcal{C}^{\infty}_{par}(\Omega)$$

whenever the latter can be defined *globally*. Indeed, assume given $g \in G$ and $U : \Delta \longrightarrow \mathbf{R}$ in (2.2), such that the mapping α in (2.11) is a $\mathcal{C}^{\infty}$-smooth diffeomorphism. In view of (3.5), we obtain $U_* \in \mathcal{C}^{\infty}_n(M)$ and then (1.5), (4.2) give

(4.5) $$\Delta \xrightarrow{\quad gU_* \quad} M$$

where

(4.6) $$(g\ U_*)(x) = g(U_*(x)) = g(x, U(x)) = (g_1(x, U(x)), g_2(x, U(x)))$$

with $x \in \Delta$. On the other hand, in view of our assumption on α, we can apply (2.12), (2.13) and obtain

(4.7) $$(g\ U)(\bar{x}) = g_2(\alpha^{-1}(\bar{x}), U(\alpha^{-1}(\bar{x}))), \text{ with } \bar{x} \in \bar{\Delta} = \alpha(\Delta)$$

therefore (3.5) gives

(4.8) $$\begin{array}{ccc} \bar{\Delta} & \xrightarrow{\quad (gU)_* \quad} & M \\ \bar{x} & \longrightarrow & (\bar{x}, g_2(\alpha^{-1}(\bar{x}), U(\alpha^{-1}(\bar{x})))) \end{array}$$

Now, from (2.13), (4.5) and (4.8) it is clear that, in general

(4.9) $$g\ U_* \neq (g\ U)_*$$

since, for instance, their domains of definition need not be the same. However, we have

(4.10) $$(g\ U_*)(\Delta) = \gamma_{gU}$$

since a direct computation gives, see (3.2), (1.2)

(4.11) $$\gamma_{gU} = g\gamma_U = \{(g_1(x, U(x)), g_2(x, U(x))) | x \in \Delta\}$$

and on the other hand, see (4.6), (1.2)

(4.12) $$(g\ U_*)(\Delta) = \{(g_1(x, U(x)), g_2(x, U(x))) | x \in \Delta\}$$

It follows that, in view of (3.3), (3.4), the Lie group action $g\ U_*$ of g on the parametric representation U_* of U, is itself a parametric representation of $g\ U$, which is the Lie group action of g on U. In other words, in general, the diagram

$$(4.13) \qquad \begin{array}{ccc} \mathcal{C}^{\infty}_{par}(\Omega) \ni U & \xrightarrow{\quad g \quad} & gU \in \mathcal{C}^{\infty}_{par}(\Omega) \\ \Big\downarrow (\ )_* & & \Big\downarrow (\ )_* \\ \mathcal{C}^{\infty}_{n}(M) \ni U_* & \xrightarrow[\quad g \quad]{} & gU_* \neq (gU)_* \in \mathcal{C}^{\infty}_{n}(M) \end{array}$$

is *not* commutative, see (4.9). Nevertheless, we have, see (3.9), (3.24), (4.11), (4.12)

$$(4.14) \qquad g\ U_* \in \mathcal{P}_{gU}, \quad g\ U_* \leq (g\ U)_*$$

Further, we note that, regardless of (4.9) and (4.13), we obtain the following *commutative* diagram

(4.15)

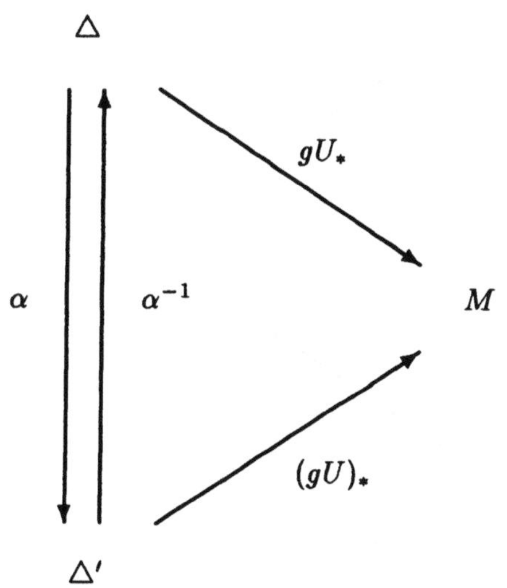

which follows easily from (4.6), (4.8) and (3.5). In this way, in view of (3.12), (3.13), the above commutative diagram means that

$$(4.16) \qquad g\ U_* \ \leq \ (g\ U)_* \ \leq \ g\ U_*$$

in other words, in case the usual Lie group action $g\ U$ of $g \in G$ on the function U exists *globally*, then its *canonical* parametric representation $(g\ U)_*$ is both *more simple* and *less simple* than $g\ U_*$, which is the Lie group action on the *canonical* parametric representation U_* of U, and which *always* exists.

Remark 1

In view of the commutative diagram (4.15), and the double inequality in (4.16), it appears to be natural to use the *global* Lie group action $g\ U_*$, which always exists, instead of the Lie group action $g\ U$, since as seen in chapter 2, the latter need not always exist.

In fact, the *essential point* in using parametric representations is that we can *abandon* $(g\ U)_*$ in (4.9), (4.13) - (4.16), and instead, use $g\ U_*$, which always exist *globally*, and which also happens to be a parametric representation of $g\ U$, whenever the latter exists globally in the classical sense, Olver [1,2].

□

Finally, related to the commutative diagram (4.15), and the double inequality (4.16), we have the following additional *universality* type properties. Given $\Lambda \xrightarrow{V} M$ a function from $\mathcal{C}_n^\infty(M)$, such that the diagram of $\mathcal{C}^\infty$-smooth mappings

(4.17)

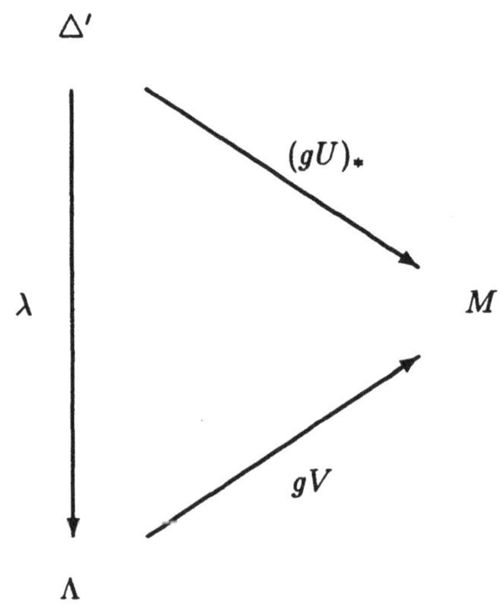

is commutative, then

$$(4.18) \quad V \circ \lambda \circ \alpha = U_*$$

Indeed, (4.15) - (4.17) yield

$$g\ U_* = g\ V \circ \lambda \circ \alpha$$

hence (4.18) follows from (4.2). Similarly, if the diagram of $\mathcal{C}^\infty$-smooth mappings

(4.19)

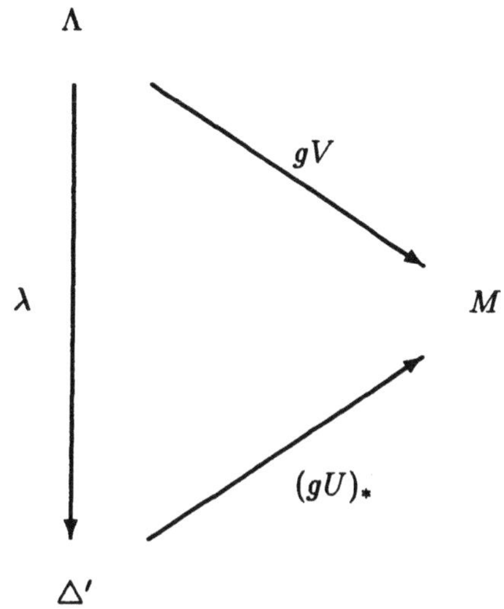

is commutative, then

(4.20) $V = U_* \circ \alpha^{-1} \circ \lambda$

Also, if the diagram of C^∞-smooth mappings

(4.21)

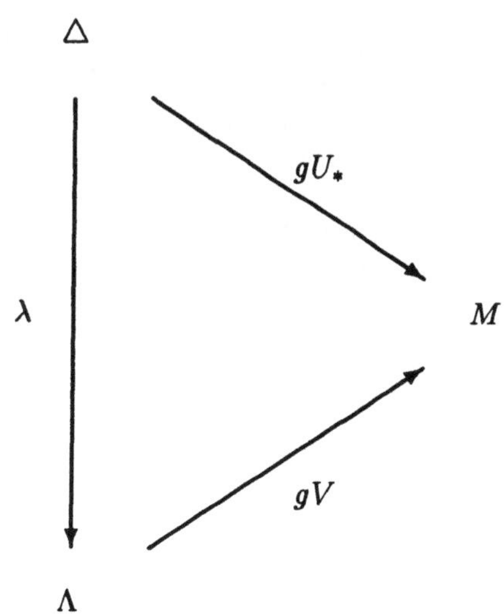

is commutative, then

(4.22) $V \circ \lambda = U_*$

Finally, if the diagram of C^∞-smooth mappings

(4.23)

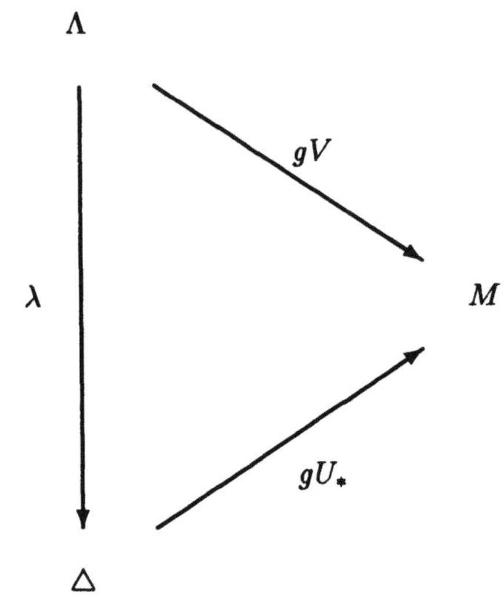

is commutative, then

$$(4.24) \quad V = U_* \circ \lambda$$

The commutative diagram (4.15), the double inequality (4.16) and the universal properties (4.17) - (4.24) give both the explanation and remedy for the failures in (4.9) and (4.13).

Comments. The *novelty* of the extension of Lie group actions on parametric functions, as defined in (4.2), when compared with the usual one in chapter 2, becomes now clear. Indeed, in the usual approach, Olver [1,2], first, one defines the Lie group action (1.2) on the manifold M of independent and dependent variables, respectively, $x \in \Omega$ and $u \in \mathbf{R}$. As a second step, one extends this initial Lie group action to functions $U : \Delta \longrightarrow \mathbf{R}$ in (2.2). This extension is done by replacing U with its graph $\gamma_U \subseteq M$, and then letting the Lie group act pointwise on γ_U, seen as a subset of M. Certainly, for every $g \in G$, we obtain in this way a well defined subset $g\gamma_U \subseteq M$. However, for *nonprojectable* Lie groups it need *not* in general happen that

$$(4.25) \quad \begin{array}{l} \exists \ \bar{U} \in \mathcal{C}^\infty_{par}(\Omega) \ : \\ g\gamma_U = \gamma_{\bar{U}} \end{array}$$

that is, the subset $g\gamma_U$ need *not* be the graph of *any* function $\bar{U} : \bar{\Delta} \longrightarrow \mathbf{R}$, where $\bar{\Delta} \subseteq \Omega$ is nonvoid, open. In this way, the usual method of extending the Lie group action (1.2) from the manifold M to functions U in $\mathcal{C}^\infty_{par}(\Omega, \mathbf{R})$, by using the graph $\gamma_U \subseteq M$ of U, is severely limited in its globality, in the case of nonprojectable Lie groups.

The way out of this nonglobality impasse, as presented in this chapter, is based on the observation that the functions U in $\mathcal{C}^\infty_{par}(\Omega)$ need *not* be seen as being defined in terms which are necessarily *internal* or *confined* to the manifold M of *independent and dependent variables* $x \in \Omega$ and $u \in \mathbf{R}$, respectively. Indeed, by introducing the *parametric representation* of functions in $\mathcal{C}^{\infty}_{par}(\Omega)$, as done in chapter 3, we can embed $\mathcal{C}^{\infty}_{par}(\Omega)$ into the space of parametric functions $\mathcal{C}^{\infty}_{n}(M)$, see (3.8). These parametric functions have *arbitrary* domains, which are nonvoid, open in $\mathbf{R}^n$, however, their *range* is *always* in the manifold M of independent and dependent variables. And as seen in (4.1) - (4.4), extending *globally* arbitrary Lie group actions (1.2) to functions in $\mathcal{C}^{\infty}_{n}(M)$ is a rather simple and straightforward procedure, being merely the *composition* of two mappings, each of which always exists. Furthermore, as seen in (4.5) - (4.24), this extension contains as a particular case the usual way Lie group actions (1.2) are extended to functions in $\mathcal{C}^{\infty}_{par}(\Omega, \mathbf{R})$, whenever these latter extensions exist globally.

For further reference, it is useful to write (4.1) - (4.4) under the stronger form: given $\Lambda \subseteq \mathbf{R}^n$ nonvoid, open, then

$$(4.26) \quad G \times \mathcal{C}^{\infty}(\Lambda, M) \longrightarrow \mathcal{C}^{\infty}(\Lambda, M)$$

Semigroup Actions. It is obvious that (4.1), (4.3) remain valid, that is, the composition gV will still be a $\mathcal{C}^{\infty}$-smooth function, and thus an element of $\mathcal{C}^{\infty}_{n}(M)$, if the mapping

$$(4.27) \quad M \xrightarrow{g} M$$

is no longer restricted to being given by the Lie group action (1.2), through the $\mathcal{C}^{\infty}$-smooth diffeomorphism (1.5), but simply, it is an arbitrary $\mathcal{C}^{\infty}$-smooth function. In other words, (4.3) actually defines the following extension of (4.1)

$$(4.28) \quad \mathcal{C}^{\infty}(M, M) \times \mathcal{C}^{\infty}_{n}(M) \longrightarrow \mathcal{C}^{\infty}_{n}(M)$$

Since $\mathcal{C}^{\infty}(M, M)$ is a noncommutative *semigroup* with identity, and not a group, when considered with the usual composition of functions, it is clear that (4.28) is a vastly larger extension of (4.1), no matter which would be the Lie group G considered.

Here, it should be noted that, lately, there has been an interest in certain *semigroup* type actions, Hilgert & Neeb, Weinstein. None of them, however, aims at the generality of (4.28). Indeed, in Hilgert & Neeb, which follows the work of the school of K H Hofmann, the Lie semigroups considered must be *subsemigroups* of Lie groups, thus they cannot include the semigroup $\mathcal{C}^{\infty}(M, M)$ which acts in (4.28). The concept of grupoid, presented in the survey of Weinstein, is similarly not capable of including the mentioned semigroups which act in (4.28).
Needless to say, the extension of the *symmetry* concept from the framework of the Lie group actions in (4.1), to that of the vastly more general semigroup actions in (4.28), can be of a significant interest in the study of PDEs, even in the case of their classical solutions.
A start in this direction is presented in chapter 13.

Chapter 5

Parametric Functions as Solutions

In chapter 4, in (4.1) - (4.4), we have seen the novel way in extending *globally* the arbitrary Lie group actions (1.2) from manifolds $M = \Omega \times \mathbf{R}$ to $\mathcal{C}^\infty$-smooth functions $U : \Delta \longrightarrow \mathbf{R}$, with $\Delta \subseteq \Omega$ nonvoid, open, by using the *parametric representation* $U_* : \Delta \longrightarrow M$ of U.
We have also seen that the respective extension works as well for functions which from the start are given parametrically.

However, in the context of PDEs, one of the major issues related to such Lie group actions is to show that, in case the Lie group action (1.2) is the symmetry group of the nonlinear PDE in (1.1), then the Lie group action upon a solution, will again give a solution.
In view of the *two track* approach mentioned at the end of chapter 3, when we prove in the sequel that the action of a Lie symmetry group on a solution of a PDE is again a solution, we shall do it for *both* the usual nonparametric, *as well as* the parametrically given versions of a function.

Detour to the Results in the Appendix. At this point, before going further, it may be useful to turn to the Appendix and become more familiar with the issue of the $\mathcal{C}^\infty$-smooth inversion of a given $\mathcal{C}^\infty$-smooth *local* diffeomorphism. In particular, the main concern here is that, through such an inversion, one single $\mathcal{C}^\infty$-smooth local diffeomorphism may typically yield a whole *family* of only *partially* defined $\mathcal{C}^\infty$-smooth diffeomorphisms. Here partial means that the *domains* of the respective functions will only be *open subsets* of the range of the initial function. Needless to say, all this is rather elementary. However, what is required in the sequel - when we go along the first track of usual, nonparametric functions - is to be able to keep in mind *all at once* the whole family of resulting partial functions which give the sought after inverse.

Back to Usual, Nonparametric Functions. Let us suppose that the $\mathcal{C}^\infty$-smooth function

$$(5.1) \quad U : \Delta \longrightarrow \mathbf{R}, \quad \Delta \subseteq \Omega, \text{ nonvoid, open}$$

is a solution of (1.1) on Δ, that is

$$(5.2) \quad T(x, D)U(x) = 0, \text{ with } x \in \Delta.$$

Further, let us suppose that the Lie group action (1.2) is a symmetry group of (1.1).

Given now $g \in G$, then as seen in (4.5), (4.6), we can always define *globally* the Lie group action

$$(5.3) \quad gU_* : \Delta \longrightarrow M$$

upon the parametric representation

$$(5.4) \quad U_* : \Delta \longrightarrow M$$

of U in (5.1). And in view of (4.16), this action gU_* is the best possible *global* replacement of the usual action gU, when the latter may fail to exist globally.

The question therefore arises : in which sense, if in any at all, is gU_* in (5.3) again a solution of the nonlinear PDE in (1.1) ?

The answer given next shows that, in a simple and natural way, gU_* is again a solution of (1.1).

Let us recall that, see (4.6).

$$(5.5) \quad (gU_*)(x) = (g_1(x, U(x)), g_2(x, U(x))), \text{ with } x \in \Delta$$

and, see (2.11).

$$(5.6) \quad \Delta \ni x \overset{\alpha}{\longmapsto} g_1(x, U(x)) \in \Omega$$

Let $J_\alpha : \Delta \longrightarrow \mathbf{R}$ be the Jacobian of α, then the set of *singularities* of α is given by

$$(5.7) \quad \textstyle\sum_{(g,U)} = \{x \in \Delta \mid J_\alpha(x) = 0\}$$

which is closed in Δ, therefore

$$(5.8) \quad \Delta_{(g,U)} = \Delta \setminus \textstyle\sum_{(g,U)} \text{ is open in } \Delta$$

Assuming that $\Delta_{(g,U)} \neq \phi$, see below (5.26) - (5.30), the inverse function theorem implies that

$$(5.9) \quad \Delta_{(g,U)} \overset{\alpha}{\longrightarrow} \bar{\Delta}_{(g,U)}$$

is a $\mathcal{C}^\infty$-smooth *surjective local diffeomorphism*, see Appendix, where

(5.10) $\bar{\Delta}_{(g,U)} = \alpha(\Delta_{(g,U)})$ is open in Ω

Let us therefore take any

(5.11) $x \in \Delta' \subseteq \Delta_{(g,U)}$

where Δ' is open and such that

(5.12) $\Delta' \xrightarrow{\alpha} \alpha(\Delta')$

is a $\mathcal{C}^\infty$-smooth diffeomorphism.

If now we restrict U to Δ', then in view of (5.12), we can apply (2.12), (2.13) and obtain the $\mathcal{C}^\infty$-smooth mapping

(5.13) $\bar{\Delta} \xrightarrow{\bar{U}} \mathbf{R}$

where $\bar{\Delta} = \alpha(\Delta')$ is open in Ω, while

(5.14) $\bar{U}(\bar{x}) = g_2(\alpha^{-1}(\bar{x}), U(\alpha^{-1}(\bar{x})))$, with $\bar{x} \in \bar{\Delta}$

But $\bar{U}$ is the result of the action of g on U when the latter is restricted to Δ', that is

(5.15) $\bar{U} = gU$ on $\bar{\Delta}$

while U satisfies (1.1) on Δ'.

Therefore, since (1.2) was supposed to be a Lie group of symmetry of (1.1), it follows that

(5.16) $T(\bar{x}, D)\bar{U}(\bar{x}) = 0, \qquad \bar{x} \in \bar{\Delta}$

Now, it only remains to establish the connection between $\bar{U}$ and gU_*, see (5.3). For that purpose we note that the conditions required for (4.15) will apply, provided that we take $\Delta = \Delta'$ in that commutative diagram. In this way, we obtain

(5.17) $gU_* = \bar{U}_* \circ \alpha \; , \bar{U}_* = gU_* \circ \alpha^{-1} \quad$ on Δ'

or pointwise

$$(5.18) \quad \begin{aligned} (gU_*)(x) &= (\alpha(x), \bar{U}(\alpha(x))), & x \in \Delta' \\ (\bar{x}, \bar{U}(\bar{x})) &= (gU_*)(\alpha^{-1}(\bar{x})), & \bar{x} \in \bar{\Delta} \end{aligned}$$

Since $M = \Omega \times \mathbf{R}$, let us denote by $pr_1 : M \longrightarrow \Omega$ and $pr_2 : M \longrightarrow \mathbf{R}$ the usual projections. Then the second relation in (5.18) gives

(5.19) $\bar{U}(\bar{x}) = pr_2(gU_*)(\alpha^{-1}(\bar{x})), \; \bar{x} \in \bar{\Delta}$

hence in view of (5.16), it follows that

(5.20) $T(\bar{x}, D)pr_2(gU_*)(\alpha^{-1}(\bar{x})) = 0, \ \bar{x} \in \bar{\Delta}$

this being the way the Lie group action gU_* on the parametric representation U_* of U, will satisfy the same PDE as U.

Finally, we should note that in view of the fact that in (5.11) we can take any $x \in \Delta_{(g,U)}$, see (A.8), it follows that (5.20) holds *locally* on $\bar{\Delta}_{(g,U)}$, see (5.10), as well as (5.56) - (5.59) below.

Advantages of Parametrisation. At this stage, we can give a short, first summary of the advantages which the parametric approach in chapters 3 - 4 brings with it, when applied to the usual Lie theory related to the study of PDEs.

First, as mentioned at the end of chapter 4, the parametric approach allows the *easy global* definition of arbitrary nonlinear Lie group actions on classical functions, by avoiding altogether the possible failure of $\mathcal{C}^\infty$-smooth invertibility of the mappings in (2.11), a failure which is typical in the case of nonprojectable Lie groups.

Second, the parametric approach preserves the essential property of Lie group symmetries, namely that solutions are transformed in solutions.
In the case of generalised solutions, the respective results will be presented in chapter 9.
Below, in Theorem 1 in this chapter, we shall see the corresponding result in the classical case, namely that, such parametric solutions are also transformed into parametric solutions, when acted upon by the Lie symmetry groups of the respective PDEs.

With respect to eventual concern felt about the less than usual, or possibly difficult nature of the parametric approach, we can mention the following.
Classical functions, in their usual nonparametric, or in their parametric form , can easily and globally be subjected to arbitrary Lie group actions, when these actions are implemented in their parametric form in (4.1) - (4.4). And the results of such Lie group actions will be classical parametric functions.
And it is only at this point, after having obtained such results, that, finally, we may have to deal with the possible lack of a $\mathcal{C}^\infty$-smooth inverse of the mappings in (2.11). And this only if we want to turn the parametrically given results of the Lie group actions back into usual, nonparametric classical functions. In such a case, we shall have to turn to the general procedure in the Appendix.

As we shall see in chapters 7 - 9, a similar situation prevails when we act upon generalised functions with arbitrary Lie group actions and we use the respective extension of the parametric form of these actions given initially in (4.1) - (4.4) for classical functions.

However, even in this case of the *first track* approach, when we want to have the results of the parametric Lie group actions obtained in the usual, nonparametric form, the following three important facts should be noted :

A) As follows from (4.1) - (4.4) and the above, the reason why the usual approach to nonprojectable Lie group actions on nonparametric functions cannot give global results is *not* related to Lie group actions, but to the fact that these actions are performed on nonparametric functions, and the end results of such actions are as well expected to be given in a nonparametric form.

B) As seen in section 7.6 and chapter 9, there is no inevitable need to turn parametrically given results of Lie group actions into usual, nonparametric ones.

Finally

C) Even when we want the final results of parametric Lie group actions to be made explicit as usual, nonparametric functions, we stil can do better and in a *less restrictive* manner than in the usual approach, Olver [1,2].

Indeed, in this usual approach, in order to be able to obtain x as a function of $\bar{x}$ from the first equation in (2.10), or equivalently, to invert in a $\mathcal{C}^\infty$-smooth way α in (2.11), one assumes locality not only with respect to x, but also with respect to $g \in G$, that is, one assumes that g is in a suitable neighbourhood of the neutral element $e \in G$. This latter assumption is particularly restrictive, as it only allows for Lie group actions which correspond to a small perturbation of the neutral element $e \in G$. Furthermore, this assumption is excessive, as it only corresponds to a most obvious and trivial technical convenience, related to the typical lack of $\mathcal{C}^\infty$-smooth invertibility of the mapping in (2.11). Namely, for $g = e$, the mapping α in (2.11) becomes

$$\Delta \ni x \overset{id_\Delta}{\longmapsto} x \in \Omega$$

which is trivially invertible. Therefore, a small perturbation of $g = e$ will still allow the *local* invertibility of α in (2.11).

On the other hand, as seen in (5.11), when we act with arbitrary Lie groups on the canonical parametrisation of usual, nonparametric functions, we shall only have to assume locality in x.
With respect to g, the main constraint results again on x, and not on g itself, namely, the initial domain Δ of the mappings U, U_*, gU_* and α is restricted to $\Delta_{(g,U)}$ in (5.8), and it is assumed that $\Delta_{(g,U)} \neq \phi$, see (5.26) - (5.30) below for further details on this condition.
A similar situation prevails in the *general* case, when arbitrary Lie groups act on any parametrically given function, see (5.60) - (5.64).

Fortunately, the above advantages of the parametric approach can be extended to the case of arbitrary Lie group actions on generalised functions, see chapters 7, 8, and especially chapter 9.

Singularity Points. Due to Sard's theorem, Kahn, a rather surprising and convenient thing happens, namely, no matter how much smaller $\Delta_{(g,U)}$ is than Δ, the size of $\bar{\Delta}_{(g,U)}$ in (5.9), (5.10) is hardly affected. Indeed, let us denote the set of *singular values* of α by, see (5.7)

(5.21) $\quad \bar{\sum}_{(g,U)} = \alpha(\sum_{(g,U)})$

then Sard's theorem gives

(5.22) $\quad \text{mes } (\bar{\sum}_{(g,U)}) = 0$

where mes denotes the Lebesgue measure on $\mathbf{R}^n$. Owing to an elementary set theoretic argument, we have

$$\alpha(\Delta) \setminus \bar{\Delta}_{(g,U)} \subseteq \bar{\sum}_{(g,U)}$$

hence (5.22) implies

(5.23) $\quad \text{mes } (\alpha(\Delta) \setminus \bar{\Delta}_{(g,U)}) = 0$

Furthermore, in view of (5.10), $\alpha(\Delta) \setminus \bar{\Delta}_{(g,U)}$ is closed in $\alpha(\Delta)$, thus (5.23) gives

(5.24) $\quad \alpha(\Delta) \setminus \bar{\Delta}_{(g,U)}$ is closed, nowhere dense in the interior of $\alpha(\Delta)$

From (5.23), (5.24) it follows that, no matter how small $\Delta_{(g,U)}$ is, it image $\bar{\Delta}_{(g,U)}$ through α will miss $\alpha(\Delta)$ by the *closed, nowhere dense* set $\alpha(\Delta) \setminus \bar{\Delta}_{(g,U)}$ which also has *zero Lebesgue measure.*

The significance of this fact is as follows. In view of (5.13) - (5.20), the set $\bar{\Delta}_{(g,U)}$ is the domain on which, *locally*, the Lie group actions gU_* and gU are related, and gU_* satisfies the PDE in (1.1). Therefore, the larger $\bar{\Delta}_{(g,U)}$, the better the results in (5.13) - (5.20). And in view of (5.23), (5.24), the open set $\bar{\Delta}_{(g,U)}$ is *maximal topologically*, being dense in $\alpha(\Delta)$, and in addition, it is *largest measurewise*, having the same Lebesgue measure with $\alpha(\Delta)$.

It should be noted that the simultaneity of (5.23) and (5.24) is *nontrivial.* Indeed, closed and nowhere dense sets in $\mathbf{R}^n$ can have arbitrarily large Lebesgue measure, Oxtoby.

Finally, we can note that, see (5.9) - (5.13)

(5.25) $\quad \cup\Delta' = \bar{\Delta}_{(g,U)}$

where the union is taken over all Δ' in (5.11), (5.12). Therefore (5.16) means that, albeit *locally*, that is, on each Δ', the Lie group action gU_* in (5.3) is again a solution of the PDE in (5.2). And in view of (5.23), (5.24), this covers the whole of $\alpha(\Delta) \subseteq \Omega$, except for the set $\alpha(\Delta) \setminus \bar{\Delta}_{(g,U)}$, which is *small* both measurewise, and topologically.

Nonsingular Points. In connection with the condition, see (5.8), (5.9)

(5.26) $\quad \Delta_{(g,U)} \neq \phi$

we note the following.

If $g = e \in G$, then (1.3), (2.11) give

$$\alpha = id_\Delta$$

hence in view of (5.8), we obtain

$$\Delta_{(g,U)} = \Delta \neq \phi$$

Therefore, owing to the continuity of α in (2.11), it follows that condition (5.26) will be satisfied for $g \in G$ which is near enough to $e \in G$. In the case of an arbitrary $g \in G$, the equation, see (5.7)

$$(5.27) \quad J_\alpha(x) = 0, \quad x \in \Delta$$

is a first order PDE in U, which in general is highly nonlinear and also depends on $g \in G$. Therefore, we shall have

$$(5.28) \quad \Delta_{(g,U)} = \phi$$

only in case (5.27) has U as a solution on the whole of Δ. However, since this solution U may depend on the particular g, it follows that, if we take $g' \in G$, $g' \neq g$, and the corresponding solution U' of (5.27), then it may easily happen that $\Delta_{(g,U')} \neq \phi$ or $\Delta_{(g',U)} \neq \phi$. Therefore, one may expect that condition (5.26) is typically satisfied. Furthermore, in case both the Lie group (1.2) and U are *analytic*, and there exists at least one $x_0 \in \Delta$ such that

$$(5.29) \quad J_\alpha(x_0) \neq 0$$

then the practically *important* property follows, Olver [2],

$$(5.30) \quad \Delta_{(g,U)} \text{ is open and dense in } \Delta$$

Derivatives of Parametric Functions. As in (3.19), let us take any parametric function $V : \Lambda \to M$ in $\mathcal{C}_n^\infty(M)$. Then $\Lambda \subseteq \mathbf{R}^n$ is nonvoid, open, and we obtain the following two $\mathcal{C}^\infty$-smooth mappings V_1, V_2, with which V is equivalent, namely

$$(5.31) \quad \begin{aligned} &\Lambda \ni y \longmapsto V(y) = (V_1(y), V_2(y)) \in M \\ &\Lambda \ni y \longmapsto V_1(y) \in \Omega \\ &\Lambda \ni y \longmapsto V_2(y) \in \mathbf{R} \end{aligned}$$

Let us consider the system of $n+1$ equations

$$(5.32) \quad \begin{aligned} x &= V_1(y) \\ U &= V_2(y) \end{aligned}$$

where $y \in \Lambda$, $x \in \Omega$, $U \in \mathbf{R}$. Let $J_{V_1} : \Lambda \longrightarrow \mathbf{R}$ be the Jacobian of V_1 in (5.31). We assume that the open set

$$(5.33) \quad D_V = \{ y \in \Lambda \mid J_{V_1}(y) \neq 0 \}$$

is nonvoid, see (5.26) - (5.30). Given now any

$$(5.34) \quad y_0 \in D_V$$

the inverse function theorem yields open neighbourhoods $\Lambda' \subseteq D_V$ and $\Delta' \subseteq \Omega$ of y_0, repsectively $x_0 = V_1(y_0)$, such that

$$(5.35) \quad V_1 : \Lambda' \longrightarrow \Delta'$$

is a $\mathcal{C}^\infty$-smooth diffeomorphism, therefore the $\mathcal{C}^\infty$-smooth inverse relation

$$(5.36) \quad y = V_1^{-1}(x), \quad x \in \Delta'$$

holds. Then using the last equation in (5.32), we obtain locally, that is, on Δ', the $\mathcal{C}^\infty$-smooth function $U : \Delta' \longrightarrow \mathbf{R}$, given by

$$(5.37) \quad U(x) = V_2(V_1^{-1}(x)), \quad x \in \Delta'$$

Here it is important to note that U in (5.37) may depend on Δ', and not only on V_1, V_2, that is V, as may be suggested by that relation. Indeed, this is the consequence of the fact that, see (5.31), the mapping $V_1 : \Lambda \longrightarrow \Omega$ need not have a global $\mathcal{C}^\infty$-smooth inverse.

The fact of interest to us, however, is the following. When we want *effectively* to compute the partial derivatives of the function U in (5.37), we can do so both *globally* and *explicitely*, straight from the original parametric form (5.31), that is, without having to go through the often only locally valid and difficult to compute steps (5.34) - (5.37). Indeed, the *chain rule* will give the relation

$$(5.38) \quad (\partial_1 U(x), ..., \partial_n U(x)) = (\partial_1 V_2(y), ..., \partial_n V_2(y))(\mathcal{J}_{V_1}(y))^{-1}$$

where, see (5.33), (5.34)

$$(5.39) \quad y \in D_V \subseteq \Lambda, \quad x = V_1(y) \in V_1(D_V) \subseteq \Omega$$

while ∂_i, with $1 \leq i \leq n$, denotes the i-th partial derivative in the respective variables. We recall here that in view of (5.33), $V_1(D_V)$ is an open subset of Ω. In (5.38), we used the notation $\mathcal{J}_W(z)$ for the differential at $z \in E$ of a mapping $W : E \longrightarrow \mathbf{R}^l$, where $E \subseteq \mathbf{R}^k$ is open. It follows that $\mathcal{J}_W(z)$ is the $l \times k$ matrix of first order partial derivatives of W, thus in particular

$$J_{V_1}(y) = \det \mathcal{J}_{V_1}(y), \quad y \in \Lambda$$

and also

$$\mathcal{J}_U(x) = (\partial_1 U(x), ..., \partial_n U(x))$$

$$\mathcal{J}_{V_2}(y) = (\partial_1 V_2(y), ..., \partial_n V_2(y))$$

with x and y as in (5.39).

Clearly, the right hand side in (5.38) can be computed *alone* in terms of the parametric representation (5.31). Furthermore, unlike (5.35) - (5.37), which in general are only valid locally, the relation (5.38) is *global*, under the condition (5.39).
Finally, *higher order* partial derivatives of U in (5.37) can also be computed *alone* in the terms of the parametric representation (5.31), simply by a further application of the chain rule to (5.38). And again, the respective relations will be valid *globally*, under the same condition (5.39), see details at the end of this chapter.

Parametric Functions as Solutions. We turn now to the second track approach, mentioned at the end of chapter 3. Namely, we shall show the following two things :

A) The way in which a *parametrically* given function, such as for instance in (5.31), is a solution of a general nonlinear PDE in (1.1). Here, as with the computation of partial derivatives of a parametrically given function, we mean both a *global* and *explicit* way of checking whether a parametric function is a solution, that is, we do so without first having to go through the typically local and difficult steps in (5.34) - (5.37).

B) The fact that, given such a parametric solution, if according to (4.3), we apply to it a Lie group action corresponding to a symmetry of the PDE in (1.1), (1.4), then we shall again *obtain* a parametric solution.

Let us note here that in (5.1) - (5.20), we have already dealt with A) and B) in the particular case, when we have a usual, nonparametric solution U, which is replaced with its *canonical* parametric representation U_*. And it should be mentioned that A) and B) are important even in such a particular case of a usual, nonparametric solution U, whenever the Lie group symmetries of the PDE in (1.1) happen to be *nonprojectable.* Indeed, given such a usual solution (5.1), (5.2), as soon as the Lie group symmetries of (5.2) are no longer projectable, one may have difficulties when trying to define *globally* their action on the given solution, see chapter 2.
Therefore, in order to overcome these difficulties, and in particular, be able to define *globally* such nonprojectable Lie group actions on usual solutions, we can now turn to the *parametric* form (5.4) of such solutions (5.1), (5.2). And then, using (4.3), we can define globally the respective Lie group actions (5.3), and we can do so without any undue restrictions whatsoever.
The results obtained, however, will again be parametric functions, and according to A), B) above, they will be solutions of the PDE in (1.1), (1.4).

Definition of Paramteric Solution. Let us turn now to presenting A) above in its general case. We assume that the nonlinear PDE in (1.1) has the form following from (1.4). Given a parametric function $V : \Lambda \longrightarrow M$ in $\mathcal{C}_n^\infty(M)$, as in (5.31), with the use of (5.32), (5.38), as well as the relations for higher order partial derivatives of U, which result from (5.38), we can define by *pull back* the nonlinear partial differential operator

$$(5.40) \quad S(y, D)\, V(y) \;=\; F(\, V_1(y),\; V_2(y),\; \ldots\,)$$

where, see (5.33), (5.39)

$$(5.41) \quad y \in D_V \subseteq \Lambda \subseteq \mathbf{R}^n$$

while in the right hand term of (5.40), the places left free and marked by dots, correspond to the partial derivatives $D_x^p U$, with $p \in \mathbf{N}^n$, $1 \leq |p| \leq m$, in (1.4). As noted, in view of (5.38), (5.39), these higher order partial derivatives of U can be computed globally in terms of the partial derivatives of V_1 and V_2 alone. Further details can be found at the end of this chapter.
Clearly, if the open subset $D_V \subseteq \Lambda$ is nonvoid, then on this set D_V, the resulting $S(y, D)$ in (5.40) is of the same form as $T(x, D)$ in (1.1), (1.4) is on Ω.

Now by definition, we shall say that the *parametric* function V in (5.31) is a *solution* of the PDE in (1.1), (1.4), if and only if the relation holds

$$(5.42) \quad S(y, D) V(y) \;=\; 0, \quad y \in D_V$$

It is important to note here that the natural *domain* on which the parametric function $V : \Lambda \to M$ is a solution, is not necessarily the whole of its domain of definition Λ, but its open subset D_V.
In view of (5.31), (5.33), we obviously have $V_1(D_V) \subseteq \Omega$, therefore in case

$$(5.43) \quad V_1(D_V) \neq \Omega$$

the resulting concept of solution given by (5.42) is weaker, that is, more general, than the usual one.
However, we should remember that the same solution may have different parametrisations. Let us, for instance, consider two such different parametrisations V, V', with V' being *simpler* than V, that is, see (3.12) - (3.14)

$$(5.44) \quad V \leq V'$$

Then the chain rule gives

$$(5.45) \quad J_{V_1}(y) \;=\; J_{V_1'}(\varphi(y))\; J_\varphi(y), \quad y \in \Lambda$$

hence, see (5.33)

$$(5.46) \quad \varphi(D_V) \subseteq D_{V'}$$

and therefore

$$(5.47) \quad V_1(D_V) \subseteq V_1'(D_{V'}) \subseteq \Omega$$

which means that, even in the case of parametric functions V, V' which are comparable in the sense of (3.12), the respective domains in (5.47) need *not* be the same.
Moreover, a parametrisation V' which is *simpler* than V, will in view of (5.47) give a *less weak* concept of parametric solution. In particular, according to (3.24), the canonical parametrisation of a solution given by a usual function will yield the least weak concept of solution, and in fact, it will give the usual concept.

Note on the Domain D_V. If $D_V = \phi$ in (5.42), then that condition is trivially satisfied. However, in such a case, obviously we cannot say that V is a parametric solution of any interest.
In order to better grasp the meaning of the concept of parametric solution introduced in (5.42), we can note the following.
First, in view of (3.31), (3.34), it is clear that there are many other parametric functions V' which are equivalent with a given parametric function V. Therefore, it is incumbent upon us to use such equivalent parametrisations V' for which $D_{V'}$ is large enough, and possibly nonvoid.
Second, such a choice is easy to make if our parametric function V is the *canonical* parametrisation of a usual function $U \in \mathcal{C}^\infty(\Delta)$, that is $V = U_*$, where $\Delta \subseteq \Omega$ is nonvoid, open. Indeed, in this case (3.5) clearly implies that

$$D_V \;=\; D_{U_*} \;=\; \Delta \;\neq \phi$$

For additional clarification on the size of the domain D_V, let us show here that the situation when

$$D_V \subset \Lambda, \quad D_V \neq \Lambda$$

can happen quite naturally, and it amounts to the *same* phenomenon which prevents even the simplest nontrivial manifolds from having one single global coordinate chart.
Indeed, let us consider the usual parametrisation of the unit circle in the plane given by

$$\Lambda = \mathbf{R} \ni y \longmapsto V(y) = (\cos y, \sin y) \in M$$

Then clearly

$$D_V = \Lambda \setminus \{\, k\pi \mid k \in \mathbf{Z} \,\}$$

therefore, one cannot choose Λ an open subset of $\mathbf{R}$, in such a way that the respective parametrisation covers the whole circle, and at the same time, we would have $D_V = \Lambda$.

Further details on the case when $D_V \neq \phi$ can be found below, in the comments related to (5.75).

Equivalent Definition of Parametric Solution. Before we turn to the proof of B) above, it will be useful to recall the typically local, however, usual nonparametric *equivalent* formulation of the parametric solution concept in (5.42). This naturally centers around (5.34) - (5.37). And in view of (5.33), (5.40), (5.41), it is clear that for $y_0 \in D_V$ and $x_0 = V_1(y_0) \in \Omega$, the relations

$$(5.48) \quad S(y, D)V(y) \; = \; 0, \quad y \in \Lambda'$$

and

$$(5.49) \quad T(x, D)U(x) \; = \; 0, \quad x \in \Delta'$$

are equivalent for suitable neighbourhoods $\Lambda' \subseteq D_V$ of y_0, respectively $\Delta' \subseteq \Omega$ of x_0, which are such that the mapping, see (5.33)

$$(5.50) \quad \Lambda' \xrightarrow{V_1} V_1(\Lambda') = \Delta'$$

is a $\mathcal{C}^\infty$-smooth diffeomorphism.

Symmetries of Parametric Solutions. We turn now to the proof of B) above. Let us assume that the parametric function V in (5.31) is a solution of the nonlinear PDE in (1.1), (1.4), namely, see (5.42)

$$(5.51) \quad S(y, D)V(y) \; = \; 0, \quad y \in D_V$$

Further, let us assume that the Lie group G in (1.2) is a symmetry group of the PDE in (1.1), (1.4). Given any $g \in G$, by using (4.3), we can always define the action gV of g on the parametric solution V.
Here we shall show that gV will again be a parametric solution of the PDE in (1.1), (1.4). Indeed, in view of (4.2), we have

$$(5.52) \quad gV \; = \; g \circ V : \Lambda \longrightarrow M$$

hence according to (5.42), we only have to try to show that

$$(5.53) \quad S(y, D)((gV)(y)) \; = \; 0, \quad y \in D_{gV}$$

For that purpose, first we note that (5.52), (5.31) and (1.3) yield

$$(5.54) \quad (gV)(y) = (g_1(V_1(y), V_2(y)), g_2(V_1(y), V_2(y))), \quad y \in \Lambda$$

hence on Λ, we shall have

$$(gV)_1 \; = \; \mathrm{pr}_2 \circ (gV) \; = \; g_1 \circ V, \quad (gV)_2 \; = \; \mathrm{pr}_2 \circ (gV) \; = \; g_2 \circ V$$

thus the chain rule gives

$$J_{(gV)_1}(y) \; = \; \det(\; \mathcal{J}_{g_1}(V(y))\mathcal{J}_V(y)), \quad y \in \Lambda$$

where as before, $\mathcal{J}_{g_1}$ and $\mathcal{J}_V$ denote the $n \times (n+1)$, respectively $(n+1) \times n$ matrices of first order partial derivatives, corresponding to $g_1 : M \longrightarrow \Omega$ and $V : \Lambda \longrightarrow M$. In this way, (5.33) results in

$$(5.55) \quad D_{gV} \; = \; \{ \, y \in \Lambda \mid \det(\, \mathcal{J}_{g_1}(V(y))\mathcal{J}_V(y)) \neq 0\}$$

We return to the proof of (5.53), and we do it by using the equivalence in (5.48) - (5.50), the assumption (5.51), and the definition (5.52).

Let us note here first that, this proof will actually be an extension of what we have already done in (5.1) - (5.20), where we were limited to the case when the parametric function (5.31) was the *canonical* parametrisation U_*, see (5.4), of a usual function U, see (5.1). It is convenient here to reformulate (5.1) - (5.20) in terms of commutative diagrams. Namely, we take any $\mathcal{C}^\infty$-smooth function

$$(5.56) \quad U : \Delta \longrightarrow \mathbf{R}$$

then clearly, we obtain the commutative diagram

$$(5.57) \qquad \begin{array}{ccccc} \Delta & \xrightarrow{\;U_*\;} & M & \xrightarrow{\;\mathrm{pr}_2\;} & \mathbf{R} \\ {\scriptstyle \mathrm{id}}\downarrow & & & & \downarrow{\scriptstyle \mathrm{id}} \\ \Delta & & \xrightarrow[\;U\;]{} & & \mathbf{R} \end{array}$$

We now apply to U_* any given Lie group action $g \in G$. For that, we take an arbitrary nonvoid open Δ', such that $\Delta' \subseteq \Delta_{(g,U)} = D_{gU_*} \subseteq \Delta \subseteq \Omega$, for which the mapping

$$(5.58) \quad \alpha = (gU_*)_1 : \Delta' \longrightarrow \alpha(\Delta') = \bar{\Delta}$$

is a $\mathcal{C}^\infty$-smooth diffeomorphism. Then we obtain the commutative diagram giving the relation $\bar{U} = gU$, that is the action of g on U, namely

$$(5.59) \qquad \begin{array}{ccccccc} \Delta' & \xrightarrow{\;U_*\;} & M & \xrightarrow{\;g\;} & M & \xrightarrow{\;\mathrm{pr}_2\;} & \mathbf{R} \\ {\scriptstyle \alpha = (gU_*)_1}\downarrow & & & & & & \downarrow{\scriptstyle \mathrm{id}} \\ \bar{\Delta} & & & \xrightarrow[\;\bar{U} = gU = \mathrm{pr}_2 \circ gU_* \circ \alpha^{-1}\;]{} & & & \mathbf{R} \end{array}$$

Furthermore, the nonvoid open subsets Δ' and $\bar{\Delta}$ in (5.58), (5.59) will cover the open sets $\Delta_{(g,U)} = D_{gU_*}$, respectively $\bar{\Delta}_{(g,U)}$, see (5.9) - (5.13).
We should note that $\bar{U}$ may depend on Δ', and not only on U and g.

Also, as we have seen in (5.1) - (5.20), if U is a solution on Δ' of the PDE in (1.1), (1.4), then $\bar{U}$ will be a solution on $\bar{\Delta}$ of that equation.

Now finally, we leave behind the particular case in (5.1) - (5.20), (5.56) - (5.59), and we turn to the general case of parametric solutions.
Let us take any V in $\mathcal{C}_n^\infty(M)$, that is

(5.60) $\quad V : \Lambda \longrightarrow M$

According to (5.34) - (5.37), let any nonvoid, open subsets $\Lambda' \subseteq D_V$ and $\Delta' \subseteq \Omega$ be such that

(5.61) $\quad V_1 : \Lambda' \longrightarrow \Delta'$

is a $\mathcal{C}^\infty$-smooth diffeomorphism. Then the commutative diagram holds

(5.62)
$$\begin{array}{ccccc}
\Lambda' & \xrightarrow{\quad V \quad} & M & \xrightarrow{\quad pr_2 \quad} & \mathbf{R} \\
\Big\downarrow V_1 & & & & \Big\downarrow \mathrm{id} \\
\Delta' & & \xrightarrow[\quad U = V_2 \circ V_1^{-1} \quad]{} & & \mathbf{R}
\end{array}$$

and the nonvoid open subsets Λ' and Δ' in (5.61), (5.62) will cover the open sets $D_V \subseteq \Lambda$, respectively $V_1(D_V) \subseteq \Omega$.
We note again that U may depend on Λ', and not only on V_1, V_2, that is, on V.

Let us apply any Lie group action $g \in G$ to the parametric function V in (5.60). Then we obtain gV as in (5.52), (5.54). Obviously, we can now apply (5.60) - (5.62) to gV. Then for any nonvoid open subsets $\Lambda' \subseteq D_{gV}$ and $\bar{\Delta} \subseteq \Omega$, such that

(5.63) $\quad (gV)_1 = g_1 \circ V : \Lambda' \longrightarrow \bar{\Delta}$

is a $\mathcal{C}^\infty$-smooth diffeomorphism, we shall obtain the commutative diagram

$$(5.64)\qquad (gV)_1 = g_1 \circ V \qquad \begin{array}{ccccccc} \Lambda' & \xrightarrow{\;V\;} & M & \xrightarrow{\;g\;} & M & \xrightarrow{\;\mathrm{pr}_2\;} & \mathbf{R} \\ \Big\downarrow & & & & & & \Big\downarrow \mathrm{id} \\ \bar{\Delta} & & & \xrightarrow[\bar{U} = (g_2 \circ V) \circ (g_1 \circ V)^{-1}]{} & & & \mathbf{R} \end{array}$$

and here again, the nonvoid open subsets Λ' and $\bar{\Delta}$ in (5.63), (5.64) will cover the open sets $D_{gV} \subseteq \Lambda$, respectively $(gV)_1(D_{gV}) \subseteq \Omega$.
Also, $\bar{U}$ may depend on Λ', in addition to V and g.

Clearly, (5.56) - (5.59) are a particular case of (5.60) - (5.64), namely, corresponding to $V = U_*$.

Before going further, we need one more preliminary step. Namely, we have to see under which condition $\bar{U}$ in (5.64) is in fact that result of the group action g on U in (5.62), that is, we have $\bar{U} = gU$, just like it happens in (5.59).
Let us take any nonvoid open subset $\Lambda' \subseteq D_V \cap D_{gV}$, for which (5.61), (5.63) hold for suitable nonvoid open subsets $\Delta', \bar{\Delta} \subseteq \Omega$. In this case both the commutative diagrams (5.62) and (5.64) will be valid. And we shall show that U and $\bar{U}$ in these diagrams do actually satisfy the commutative diagram (5.59) as well, in other words, we have

$$(5.65)\qquad \bar{U} = gU \quad \text{on } \bar{\Delta}$$

Indeed, (5.62), (5.64) give

$$(5.66)\qquad U = V_2 \circ V_1^{-1} \quad \text{on } \Delta', \qquad \bar{U} = (g_2 \circ V) \circ (g_1 \circ V)^{-1} \quad \text{on } \bar{\Delta}$$

Further, (5.61), (5.63) imply that

$$(5.67)\qquad g_1 \circ V \circ V_1^{-1} : \Delta' \longrightarrow \bar{\Delta}$$

is a C^∞-smooth diffeomorphism. But (5.66) results in

$$(5.68)\qquad (gU_*)_1 = g_1 \circ V \circ V_1^{-1} \quad \text{on } \Delta'$$

since for $x \in \Delta'$, we have

$$(gU_*)_1(x) = g_1(x, U(x))$$

$$(g_1 \circ V \circ V_1^{-1})(x) = g_1(V(V_1^{-1}(x))) = g_1(V_1(V_1^{-1}(x)), V(V_1^{-1}(x))) = g_1(x, U(x))$$

Also

$$(5.69)\qquad (g_2 \circ V) \circ (g_1 \circ V)^{-1} = \mathrm{pr}_2 \circ gU_* \circ (gU_*)_1^{-1} \quad \text{on } \bar{\Delta}$$

as for $\bar{x} \in \bar{\Delta}$, we obtain

$$(g_1 \circ V)^{-1}(\bar{x}) = y \in \Lambda'$$

$$(g_1 \circ V)(y) = g_1(V(y)) = \bar{x}$$

$$((g_2 \circ V) \circ (g_1 \circ V)^{-1})(\bar{x}) = (g_2 \circ V)(y) = g_2(V(y))$$

while (5.68) gives

$$(gU_*)_1^{-1} = (g_1 \circ V \circ V_1^{-1})^{-1} = V_1 \circ (g_1 \circ V)^{-1}$$

hence

$$(gU_*)_1^{-1}(\bar{x}) = V_1 \circ (g_1 \circ V)^{-1}(\bar{x}) = V_1(y)$$

$$\begin{aligned}(\mathrm{pr}_2 \circ gU_*)((gU_*)_1^{-1}(\bar{x})) &= (g_2 \circ (V_2 \circ V_1^{-1})_*)(V_1(y)) \\ = g_2(V_1(y), V_2(V_1^{-1}(V_1(y)))) &= g_2(V_1(y), V_2(y)) = g_2(V(y))\end{aligned}$$

And clearly, (5.66) - (5.69) show that indeed $U, \bar{U}$ satisfy the commutative diagram (5.59), and in particular, the relation (5.65).

Now let us assume that V in (5.60) is a solution of the PDE in (1.1), (1.4). Then (5.61) - (5.64) and the equivalence in (5.48) - (5.50) means that for every commutative diagram (5.62), we obtain

(5.70) $\quad T(x, D)U(x) \;=\; 0, \quad x \in \Delta'$

and the nonvoid open subsets Δ' will cover the open set $V_1(D_V)$.
Let us restrict the family of commutative diagrams (5.62) to those only for which $\Lambda' \subseteq D_V \cap D_{gV}$. In this case (5.65) gives

(5.71) $\quad T(\bar{x}, D)\bar{U}(\bar{x}) \;=\; 0, \quad \bar{x} \in \bar{\Delta}$

and the nonvoid open subsets $\bar{\Delta}$ will cover the open set $(gV)_1(D_V \cap D_{gV})$.

Returning to the equivalence in (5.48) - (5.50), it follows that (5.71) together with (5.64) imply

(5.72) $\quad S(y, D)(gV)(y) \;=\; 0, \quad y \in \Lambda'$

with the nonvoid open subsets Λ' covering the open set $D_V \cap D_{gV}$.

Here it is important to note that, in view of (5.40), (5.41), the left hand term in (5.72) is defined *globally* on D_{gV}. Therefore finally, (5.72) will give the following weaker version of (5.53)

(5.73) $\quad S(y, D)(gV)(y) \;=\; 0, \quad y \in D_V \cap D_{gV}$

We can now summarize the above in

Theorem 1

Let $V : \Lambda \longrightarrow M$ be a parametric solution on the nonvoid open set $D_V \subseteq \Lambda$ of the nonlinear PDE in (1.1), (1.4), and let G be the Lie group of symmetries of that equation.
Given any $g \in G$, if

$$(5.74) \quad D_V \cap D_{gV} \neq \phi$$

then the parametric function $gV : \Lambda \longrightarrow M$, which is the result of the action of g on V, will again be a parametric solution of the PDE in (1.1), (1.4) on the nonvoid open set $D_V \cap D_{gV}$.

The Nature of the Domain Condition (5.74). At first sight, it may appear that the validity of Theorem 1 above would unduly be restricted by the *domain* condition (5.74). However, we can immediately note that such is not the case, and that the respective domain condition is natural, as well as minimal. Indeed, (5.74) *only* asks that the domain on which *both* V and its Lie group transformation gV are nonsingular, should be nonvoid. And in view of the commutative diagrams (5.62), (5.64), this domain condition (5.74) appears indeed to be necessary, since as seen in (4.1) - (4.4), both a parametric function V and any of its Lie group transformations gV, will by necessity have the *same* domain of parameters.
More importantly yet, in many applications, this domain condition (5.74) is actually satisfied in a much *stronger* form.
Indeed, as we noted, according to (5.33), the open subsets D_V and D_{gV} are respectively the sets on which both V and its Lie group transformation gV are not singular. In particular, in case V happens to be the canonical parametric representation of an analytic function U in (5.1), and also, the Lie group is analytic, then (5.30) implies that both D_V and D_{gV} are open and *dense* in $\Lambda = \Delta$. In this case, instead of (5.74), we shall actually have satisfied the *much stronger* condition

$$(5.75) \quad D_V \cap D_{gV} \quad \text{open and dense in} \quad \Lambda = \Delta$$

In the general case of a parametrically given function V, which need not be the parametric representation of any usual function U in (5.1), it is clear that V can be given by many parametrisations which are *equivalent* in the sense of (3.31), (3.32). In particular, if any two parametrisations V and V' are comparable, say, $V \leq V'$, see (3.12) - (3.14), then according to (3.34), they are equivalent. And as seen in (5.44) - (5.47) for instance, the respective domains may vary. Therefore, one is supposed to use a better equivalent parametrisation, in order to try to obtain the domain condition (5.74). And in this case, better will mean *simpler*, see Example 3, in chapter 3.

Let us give a very simple and rather natural example of a Lie group action which shows that we can have the situation when

$$(5.76) \quad D_V = \Lambda, \quad D_{gV} \subset \Lambda, \quad D_{gV} \neq \Lambda$$

Indeed, let us take

$$\Omega = \mathbf{R} \ni x \longmapsto U(x) = x^2 \in \mathbf{R}$$

then its canonical parametrisation U_* is

$$\Lambda = \Omega = \mathbf{R} \ni y = x \longmapsto V(y) = (y, y^2) = U_*(x) \in M = \mathbf{R}^2$$

therefore $D_V = \Lambda$ and the first part of (5.76) holds.
Now we apply to $V = U_*$ the nonprojectable Lie group action of the rotation of the plane in Example 2, in chapter 3. Let us therefore take a fixed $\theta = g \in G = (\mathbf{R}, +)$, then

$$(\theta\ U)(x) = (x \cos\theta - x^2 \sin\theta, x \sin\theta + x^2 \cos\theta), \quad x \in \Omega = \mathbf{R}$$

therefore, see (2.11)

$$\mathbf{R} \ni x \longmapsto \alpha(x) = x \cos\theta - x^2 \sin\theta \in \mathbf{R}$$

$$J_\alpha(x) = \cos\theta - 2x \sin\theta, \quad x \in \mathbf{R}$$

Now if θ is such that $\sin\theta \neq 0$, for instance, $\theta = \pi/4$, then clearly

$$J_\alpha(x) \neq 0 \iff x \neq 1/2$$

which means that $D_{gV} = \mathbf{R} \setminus \{1/2\}$, and the second part of (5.76) also holds.

Contact Transformations. Let us look somewhat further into the consequences of (5.38), (5.39), for *higher order* partial derivatives of parametric functions.
As already mentioned, the relations (5.38), (5.39) allow an *explicit* and *global* computation of all higher order partial derivatives of the *family* of usual, nonparametric, and typically only *local* functions U in (5.37), which are associated with the given parametric function V in (5.31), see also Appendix for details of that association. Indeed, the higher order partial derivatives of such functions U will be given by the successive application to the appropriate terms in (5.38) of the respective amount of first order partial derivatives. In this way, we obtain

$$(5.77) \quad D_x^p U(x) = R^p(\ldots, D_y^r V_1(y), \ldots, D_y^s V_2(y), \ldots), \quad p \in \mathbf{N}^n$$

with $y \in D_V, x = V_1(y) \in V_1(D_V) \subseteq \Omega$, and where $r, s \in \mathbf{N}^n$, $r, s \leq p$, while R^p are *rational* functions in their respective arguments.
For brevity, we denote by $V^p(y)$ the set of the arguments of R^p.

It will be convenient to use here the jet space concept of Ehresmann, see Olver [1,2] for notation. And as before, for simplicity, we shall stay with the case of one single, real valued function, of n real variables. Thus, for any $m \in \mathbf{N}$, $m \geq 0$, we have the m-th order jet space over M, given by

(5.78) $\quad M^{(m)} \;=\; \Omega \times U^{(m)}$

where $U^{(m)} = U_0 \times U_1 \times \ldots \times U_m$, while $U_i = \mathbf{R}^{n_i}$, with

$$n_i \;=\; \binom{n+i-1}{i}, \qquad 0 \leq i \leq m$$

Given $f \in C^\infty(\Omega, \mathbf{R})$, one defines the m-th *prolongation* of f as being the C^∞-smooth mapping

(5.79) $\quad f^{(m)} : \Omega \ni x \longmapsto (\ldots, D^p_x f(x), \ldots) \in U^{(m)}$

where $p \in \mathbf{N}^n, \; |p| \leq m$.

Now, in view of (5.77), see also (5.31) - (5.39), we can extend to *parametric* functions the above concept of prolongation of usual functions in (5.79). Indeed, given a parametric function $V : \Lambda \longrightarrow M$ in $C^\infty_n(M)$, we shall define its m-th prolongation

(5.80) $\quad V^{(m)} : D_V \longrightarrow M^{(m)}$

by

(5.81) $\quad V^{(m)}(y) = (V_1(y), V_2(y), \ldots, R^p(V^p(y)), \ldots), \quad y \in D_V$

where $p \in \mathbf{N}^n, \; 1 \leq |p| \leq m$. It follows that the m-th prolongation $V^{(m)}$ of a parametric function V is again a paramteric function, and its domain of parameters $D_V \subseteq \Lambda$, is the set of *nonsingular* points of V.

Let us now recall that the *contact* tranformations on M are given by certain particular C^∞-smooth mappings of the various $M^{(m)}$ into themselves, where $m \in \mathbf{N}, \; m \geq 1$, namely

(5.82) $\quad T^{(m)} : M^{(m)} \longrightarrow M^{(m)}$

mappings which are supposed to satisfy the so called contact conditions, see for details Olver [1,2].

Let us now see the way such a contact transformation (5.82) can be made to act on a parametric function V, that is, how we can define the result $T^{(m)}V$ of its action. Clearly, the m-th prolongation $V^{(m)}$ of V, see (5.80), can easily and naturally be acted upon by the contact transformation $T^{(m)}$ in (5.82), namely, by the composition of C^∞-smooth mappings

(5.83) $\quad T^{(m)}V = T^{(m)} \circ V^{(m)} : D_V \longrightarrow M^{(m)}$

and the result will be a parametric function whose domain of parameters $D_V \subseteq \Lambda$ is again the set of nonsigular points of the parametric function V, upon which the contact transformation $T^{(m)}$ has acted. Therefore, both $T^{(m)}V$ and $V^{(m)}$ have the *same* domain D_V of definition.
Here we can note that the definition of the action in (5.83) remains valid, even if $T^{(m)}$ in (5.82) is *not* necessarily a contact transformation.

Chapter 6

Rarefaction Waves and Riemann Solvers of the Nonlinear Shock Wave Equation

This chapter is introductory to chapter 8, and it only aims to illustrate in the case of the rarefaction waves and Riemann solvers of the nonlinear shock wave equation some of the approaches presented earlier in chapters 1 - 5. This prepares the way for the full treatment of applications presented in chapters 8 and 9.
The interest in these applications comes from the following two facts :

a) the rarefaction waves and Riemann solvers are *nonclassical* solutions,
b) the nonlinear shock wave equation has *nonprojectable* Lie group symmetries.

The full approach to the way the nonprojectable Lie group symmetries of the nonlinear shock wave equation act on the rarefaction waves and the Riemann solvers is presented in chapter 8. This approach - when going along the traditional first track which uses nonparametric functions - will need some of the general and technically rather involved theory in chapter 7.
On the other hand, a much simpler approach - along the second track of parametric functions - will be presented in chapter 9.

Part of this general theory in chapter 7, which is of interest to both tracks, that is, the usual and the parametric ones, is the *core* of this work, as it shows the way arbitrary nonlinear, including nonprojectable, Lie group actions can be defined on the generalised functions in the nowhere dense differential algebras (1.6).
And by comparing the approaches and results in chapters 7, 8 and 9, one can obtain a good insight into their particular advantages and disadvantages.

Here in this chapter, however, we shall only deal with the nonprojectable Lie group symmetries of the nonlinear shock wave equation, when they act on suitable, clas-

sical restrictions of the rarefaction waves and Riemann solvers. More precisely, these restrictions are on open dense subsets of the domains of the mentioned non-classical solutions, those largest such subsets on which, by avoiding their respective singularities, they happen to be classical solutions, see (6.11).
Let us consider the nonlinear shock wave equation

$$(6.1) \quad U_t + UU_x = 0, \quad t \geq 0,, \quad x \in \mathbf{R}$$

with the initial condition

$$(6.2) \quad U(0,x) = u_\ell + (u_r - u_\ell)H(x), \quad x \in \mathbf{R}$$

Here, as is well known, Smoller, there are two distinct nontrivial cases, namely

$$(6.3) \quad u_l < u_r$$

and respectively

$$(6.4) \quad u_l > u_r$$

The case (6.3) leads to the rarefaction wave solutions of the nonlinear shock wave equation, given by

$$(6.5) \quad U(t,x) = \left| \begin{array}{ll} u_l & \text{if } x \leq u_l t \\ \\ x/t & \text{if } u_l t \leq x \leq u_r t \\ \\ u_r & \text{if } u_r t \leq x \end{array} \right.$$

with $t > 0$, $x \in \mathbf{R}$, or in a form which will be convenient later

$$(6.6) \quad \begin{aligned} & U(t,x) = \\ & \quad = u_l H(u_l t - x) + (x/t)H(x - u_l t)H(u_r t - x) + u_r H(x - u_r t) = \\ & \quad = u_l + (1/t)((x - u_l t)_+ - (x - u_r t)_+), \quad t > 0, x \in \mathbf{R} \end{aligned}$$

while the case (6.4) corresponds to the Riemann solver

$$(6.7) \quad U(t,x) = \left| \begin{array}{ll} u_l & \text{if } x < st \\ \\ u_r & \text{if } st < x \end{array} \right.$$

also with $t > 0$, $x \in \mathbf{R}$, and which can be written in the useful form

$$(6.8) \quad U(t,x) = u_l + (u_r - u_l)H(x - st), \quad t > 0, x \in \mathbf{R}$$

Here $s = (u_l + u_r)/2$. Also, we denoted by $H : \mathbf{R} \longrightarrow \mathbf{R}$ the Heaviside function, which is such that $H(x) = 0$ for $x < 0$, and $H(x) = 1$ for $x > 0$. The function $(x)_+ : \mathbf{R} \longrightarrow \mathbf{R}$ is the primitive of H, hence $(x)_+ = 0$, for $x \leq 0$, and $(x)_+ = x$, for $x \geq 0$.

As is known, Ibragimov, one of the *nonprojectable* Lie group symmetries of the nonlinear shock wave equation (6.1) is given by

$$(6.9)\qquad \begin{aligned} \bar{t} &= t \\ \bar{x} &= x + \epsilon a(u) \\ \bar{u} &= u \end{aligned}$$

where $\epsilon \in \mathbf{R}$ is the additive group parameter, while $a : \mathbf{R} \longrightarrow \mathbf{R}$ is any $\mathcal{C}^\infty$-smooth function. In terms of (1.2), we shall then have

$$\Omega = (0, \infty) \times \mathbf{R}, \;\; M = \Omega \times \mathbf{R} = (0, \infty) \times \mathbf{R}^2, \;\text{ and } G = (\mathbf{R}, +)$$

while for $(t, x) \in \Omega, u \in \mathbf{R}, \epsilon = g \in G = \mathbf{R}$, we have

$$(6.10)\qquad \begin{aligned} g_1(t, x, u) &= (t, x + \epsilon a(u)) \\ g_2(t, x, u) &= u \end{aligned}$$

The rarefaction wave (6.5) and the Riemann solver (6.7) are not $\mathcal{C}^\infty$-smooth on Ω, although the former is continuous. Let us denote their respective singularity sets, and their complementaries in Ω, by

$$(6.11)\qquad \begin{aligned} &\Gamma_F = \{(t, x) \in \Omega \mid x = u_l t \text{ or } x = u_r t\}, \quad \Delta_F = \Omega \setminus \Gamma_F \\ &\Gamma_R = \{(t, x) \in \Omega \mid x = st\}, \quad \Delta_R = \Omega \setminus \Gamma_R \end{aligned}$$

Then clearly, the restricted rarefaction wave and Riemann solver are $\mathcal{C}^\infty$-smooth on Δ_F and Δ_R, respectively. Thus, in this case, the mapping (2.11) becomes for $\Delta = \Delta_F$, respectively $\Delta = \Delta_R$

$$(6.12)\qquad \Delta \ni (t, x) \stackrel{\alpha}{\longmapsto} (t, x + \epsilon a(U(t, x))) \in (0, \infty) \times \mathbf{R}$$

which for both the rarefaction wave (6.5) and the Riemann solver (6.7) will in general fail to be injective.
However, we note that, owing to (6.9), the mapping α is an identity in its first coordinate. Therefore, it will have a $\mathcal{C}^\infty$-smooth inverse, if and only if it does so in its second coordinate. In particular, its Jacobian is nonvanishing on Δ, if and only if

$$(6.13)\qquad \alpha_x(t, x) = 1 + \epsilon a'(U(t, x))U_x(t, x) \neq 0, \quad (t, x) \in \Delta$$

In other words, in order to see whether α in (6.12) has a $\mathcal{C}^\infty$-smooth inverse, it suffices to check it for an arbitrary but fixed $t > 0$, and with α considered as a function only of x, with $(t, x) \in \Delta$. Clearly, this is a simplification which is due to the particular nature of the nonprojectable Lie symmetry group (6.9) of the nonlinear shock wave equation (6.1).

Let us now turn for the moment to the *parametric* representation of the *restricted* rarefaction wave and the Riemann solver. Then, according to (3.5), we can take for instance

$$(6.14) \quad \Delta \ni (t,x) \overset{U_*}{\longmapsto} (t,x,U(t,x)) \in M$$

as one of the parametric representations. And in fact, in the sense of (3.12) - (3.28), it follows that U_* in (6.14) is the *canonical* parametric representation of U.

Given now $\epsilon = g \in G = \mathbf{R}$, by using (4.2), (6.10), we can *easily* define the nonprojectable Lie group action

$$(6.15) \quad \begin{array}{c} \Delta \overset{gU_*}{\longrightarrow} M \\ (t,x) \longmapsto (t, x + \epsilon a(U(t,x)), U(t,x)) \end{array}$$

where Δ is Δ_F, Δ_R, respectively, and thus obtain the *parametric* result gU_*. However, if we want to revert to the traditional, nonparametric form of gU_*, we shall again face the same difficulties, in view of the possible lack of invertibility of the mapping α in (6.12).

For the clarification of the difficulties involved, when U is a rarefaction wave (6.5) or the Riemann solver (6.7), it will be convenient to partition the respective domains in (6.11), as follows

$$(6.16) \quad \begin{array}{l} \Delta_F = \Delta_{Fl} \cup \Delta_{Fc} \cup \Delta_{Fr} \\ \\ \Delta_R = \Delta_{Rl} \cup \Delta_{Rr} \end{array}$$

where $\Delta_{Fl} = \{(t,x) \in \Omega \mid x < u_l t\}$, $\Delta_{Fc} = \{(t,x) \in \Omega \mid u_l t < x < u_r t\}$, $\Delta_{Fr} = \{(t,x) \in \Omega \mid u_r t < x\}$, while $\Delta_{Rl} = \{(t,x) \in \Omega \mid x < st\}$, $\Delta_{Rr} = \{(t,x) \in \Omega \mid st < x\}$.

We note that for the rarefaction wave (6.5), the Jacobian of α in (6.12) is

$$(6.17) \quad J_\alpha(t,x) = \left| \begin{array}{l} 1 \quad \text{if} \quad (t,x) \in \Delta_{Fl} \\ 1 + \epsilon a'(x/t)/t \quad \text{if} \quad (t,x) \in \Delta_{Fc} \\ 1 \quad \text{if} \quad (t,x) \in \Delta_{Fr} \end{array} \right.$$

while for the Riemann solver (6.7), it is

$$(6.18) \quad J_\alpha(t,x) = 1, \quad (t,x) \in \Delta_R$$

In this way, for the rarefaction wave, the relations (5.7) - (5.10) become

$$(6.19) \quad \begin{array}{l} \Sigma_{F,(g,U)} = \{(t,x) \in \Delta_{Fc} \mid t + \epsilon a'(x/t) = 0\} \\ \\ \Delta_{F,(g,U)} = \Delta_{Fl} \cup (\Delta_{Fc} \setminus \Sigma_{F,(g,U)}) \cup \Delta_{Fr} \\ \\ \bar{\Sigma}_{F,(g,U)} = \alpha(\Sigma_{F,(g,U)}), \quad \bar{\Delta}_{F,(g,U)} = \alpha(\Delta_{F,(g,U)}) \end{array}$$

and we have the following $\mathcal{C}^\infty$-smooth diffeomorphisms, see (6.5), (6.16)

$$(6.20) \quad \Delta_{Fl} \overset{\alpha}{\longmapsto} \alpha(\Delta_{Fl}), \quad \Delta_{Fr} \overset{\alpha}{\longmapsto} \alpha(\Delta_{Fr})$$

For the Riemann solver the relations (5.7) - (5.10) are

$$(6.21)\qquad \begin{aligned}&\Sigma_{R,(g,U)} = \phi, \ \Delta_{R,(g,U)} = \Delta_R \\ &\bar{\Sigma}_{(g,U)} = \phi, \ \bar{\Delta}_{R,(g,U)} = \alpha(\Delta_R)\end{aligned}$$

and (6.7), (6.16) give the $\mathcal{C}^\infty$-smooth diffeomorphisms

$$(6.22)\qquad \Delta_{Rl} \overset{\alpha}{\longmapsto} \alpha(\Delta_{Rl}), \quad \Delta_{Rr} \overset{\alpha}{\longmapsto} \alpha(\Delta_{Rr})$$

Let us now try to find the largest open subsets in (6.16), to which the respective $\mathcal{C}^\infty$-smooth diffeomorphisms (6.20), (6.22) can be extended.
For that, we shall deal separately with the two cases.

For the rarefaction wave (6.5), we obtain from (5.9), (6.19), that

$$(6.23)\qquad \Delta_{F,(g,U)} \overset{\alpha}{\longrightarrow} \alpha(\Delta_{F,(g,U)})$$

is a $\mathcal{C}^\infty$-smooth surjective local diffeomorphism. Let us first try to determine how large is $\Delta_{F,(g,U)}$. In view of (6.19), (6.13), let us denote

$$(6.24)\qquad \Omega_g \ = \ \{(t,x) \in \Omega \mid t + \epsilon a'(x/t) > 0\}$$

then (6.17) gives

$$(6.25)\qquad \Delta_{F,(g,U)} \supseteq \Delta_{Fl} \cup (\Delta_{Fc} \cap \Omega_g) \cup \Delta_{Fr}$$

Now by denoting $\epsilon T = -\inf\{\epsilon a'(y) \mid y \in [u_l, u_r]\}$, it follows from (6.24) that

$$(6.26)\qquad \Omega_g \supseteq \Omega_g^0 \ = \ \{(t,x) \in \Omega \mid t > \epsilon T\}$$

Furthermore, if

$$(6.27)\qquad \epsilon a'(y) \geq 0, \quad y \in [u_l, u_r]$$

then we obtain from (6.17), (6.19), (6.24) and (6.26) that

$$(6.28)\qquad \begin{aligned}&\Omega_g^0 \ = \ \Omega_g \ = \ \Omega \\ &\Sigma_{F,(g,U)} = \phi, \ \Delta_{F,(g,U)} \ = \ \Delta_F\end{aligned}$$

and in view of (6.13), the relation (6.23) obtains its strongest possible form

$$(6.29)\qquad \Delta_F \overset{\alpha}{\longrightarrow} \alpha(\Delta_F)$$

is a $\mathcal{C}^\infty$-smooth diffeomorphism.

In the general case, however, (6.5), (6.13), (6.24) and (6.25) will still imply that (6.23) is valid in the partly stronger form, namely

$$(6.30)\qquad \Delta_{F,g} \overset{\alpha}{\longrightarrow} \alpha(\Delta_{F,g})$$

is a $\mathcal{C}^\infty$-smooth diffeomorphism, where we denoted

$$(6.31)\qquad \Delta_{F,g} \ = \ \Delta_F \cap \Omega_g^0$$

In this way we can apply (2.12), (2.13) and (4.15), and obtain

(6.32) $\bar{U} = gU \ : \ \bar{\Delta} = \alpha(\Delta_{F,g}) \longrightarrow \mathbf{R}$

where

$$(6.33) \quad \bar{U}(\bar{t},\bar{x}) = \left| \begin{array}{l} u_l \ \text{ if } \ (\bar{t},\bar{x}) \in \alpha(\Delta_{Fl} \cap \Omega_g^0) \\ x/t \ \text{ if } \ (\bar{t},\bar{x}) \in \alpha(\Delta_{Fc} \cap \Omega_g^0) \\ u_r \ \text{ if } \ (\bar{t},\bar{x}) \in \alpha(\Delta_{Fr} \cap \Omega_g^0) \end{array} \right.$$

and in view of (6.12), (6.16), (6.24) and (6.25), we have the open sets

$$(6.34) \quad \begin{array}{rcl} \alpha(\Delta_{Fl} \cap \Omega_g^0) & = & \{(t, x + \epsilon a(u_l)) \mid (t,x) \in \Omega, \ t > \epsilon T, \ x < u_l t\} \\ \alpha(\Delta_{Fc} \cap \Omega_g^0) & = & \{(t, x + \epsilon a(x/t)) \mid (t,x) \in \Omega, \ t > \epsilon T, \ u_l t < x < u_r t\} \\ \alpha(\Delta_{Fr} \cap \Omega_g^0) & = & \{(t, x + \epsilon a(u_r)) \mid (t,x) \in \Omega, \ t > \epsilon T, \ u_r t < x\} \end{array}$$

Here it is important to note that the *nonprojectability* of the Lie group action (6.9) does not prevent α in (6.30) from being $\mathcal{C}^\infty$-smooth invertible. Applying then the usual theory on the nonvoid open subset $\alpha(\Delta_{F,g}) \subseteq \Omega$, we obtain $\bar{U} = gU$ in (6.31), (6.32), which is obviously again a solution, namely, in the traditional *nonparametric* form, of the nonlinear shock wave equation (6.1) on the mentioned domain.

We return now to the Riemann solver (6.7), then (6.12) gives the open sets

$$(6.35) \quad \begin{array}{rcl} \alpha(\Delta_{Rl}) & = & \{(t, x + \epsilon a(u_l)) \mid (t,x) \in \Delta_{Rl}\} \\ \alpha(\Delta_{Rr}) & = & \{(t, x + \epsilon a(u_r)) \mid (t,x) \in \Delta_{Rr}\} \end{array}$$

In this way, there are two cases to consider, namely, when

(6.36) $\alpha(\Delta_{Rl}) \cap \alpha(\Delta_{Rr}) = \phi$

and on the other hand, when

(6.37) $\alpha(\Delta_{Rl}) \cap \alpha(\Delta_{Rr}) \neq \phi$

In view of (6.35), it is clear that (6.36) is equivalent with

(6.38) $\epsilon a(u_l) \leq \epsilon a(u_r)$

while (6.37) is equivalent with

(6.39) $\epsilon a(u_r) < \epsilon a(u_l)$

In the case of (6.36), (6.38), the relation (6.22) implies that, see (6.12)

(6.40) $\Delta_R \xrightarrow{\alpha} \alpha(\Delta_R)$

is a $\mathcal{C}^\infty$-smooth diffeomorphism. Therefore, (2.12), (2.13) and (4.15) apply and we obtain

(6.41) $\bar{U} = gU : \bar{\Delta}_R = \alpha(\Delta_R) \longrightarrow \mathbf{R}$

where

$$(6.42) \quad \bar{U}(\bar{t}, \bar{x}) = \left| \begin{array}{ll} u_l & \text{if } (\bar{t}, \bar{x}) \in \alpha(\Delta_{Rl}) \\ & \\ u_r & \text{if } (\bar{t}, \bar{x}) \in \alpha(\Delta_{Rr}) \end{array} \right.$$

In this case again, the *nonprojectability* of the Lie group action (6.9) does *not* prevent α in (6.12) from being $\mathcal{C}^\infty$-smooth invertible, therefore, the usual theory can be used, and $\bar{U} = gU$ in (6.41), (6.42) is clearly a traditional *nonparametric* solution of the shock wave equation (6.1) on the open domain $\alpha(\Delta_R) \subseteq \Omega$.

In the case (6.37), (6.39), in view of (6.35), it is clear that

$$\alpha(\Delta_{Rl}) = \{(t, x) \mid x < st + \epsilon a(u_l)\}$$

$$(6.43) \quad \alpha(\Delta_{Rr}) = \{(t, x) \mid st + \epsilon a(u_r) < x\}$$

$$\alpha(\Delta_{Rl}) \cap \alpha(\Delta_{Rr}) = \{(t, x) \mid st + \epsilon a(u_r) < x < st + \epsilon a(u_\ell)\} \neq \phi$$

therefore α in (6.12) will *no longer* be invertible, let alone $\mathcal{C}^\infty$-smooth invertible, hence (6.41), (6.42) are this time out of reach.

Chapter 7

Arbitrary Nonlinear Lie Group Actions on Generalised Functions

As mentioned, a main aim of this work is to define arbitrary nonlinear, including nonprojectable, Lie group actions on generalised functions, and then, to apply such actions - when they correspond to usual Lie group symmetries of given nonlinear PDEs - to the study of the generalised solutions of the respective equations.
In this regard, one the main results obtained is that such Lie group actions on generalised solutions will again give generalised solutions, see chapters 8 and 9. This result extends the similar one for classical solutions, which is one of the cornerstones of Lie's original program.

Much of this chapter, in an attempt to go along *tradition*, is dedicated to the first track approach mentioned at the end of chapter 3, namely, it uses the customary, *nonparametric* representation of generalised functions.
This approach, however, proves to be rather involed technically, especially starting with section 7.4.
The second track approach, based on parametric representation of generalised functions will be presented in chapter 9, see also (7.3.12) - (7.3.15), (7.4.64) - (7.4.66), and section 7.6 in this chapter.
There, in chapter 9, the comparative simplicity will become quite obvious. However, even in the case of the traditional approach in this chapter, one need not always go up to the end of the ultimate technical complication, see the mentioned section 7.6.

Based on the constructions in chapters 3, 4 and 5, and with the help of a temporary *detour* into parametric representation, as suggested in (7.3.12) - (7.3.15), it proves to be surprisingly *esay* to define arbitrary, including nonprojectable, nonlinear Lie group actions on the generalised functions which are elements of the nowhere dense differential algebras A in (1.6).

Indeed, it is *important* to point out that the only technical difficulties encountered will *not at all* be related to Lie groups, or to their actions on generalised functions, but solely to the well known *classical* problem which we already have to face, Olver [1,2], when trying to define a *nonprojectable* Lie group action just on one single $\mathcal{C}^\infty$-smooth function which is given in a nonparametric form. Namely, the only problem will be related to the *return* in (7.3.15) from the detour in (7.3.13).

The problem which we have to deal with in such a situation has nothing to do with Lie group actions, and it is but the classical calculus one of inverting locally a surjective local diffeomorphism, which fails to be a diffeomorphism. This problem, in its full generality, is solved in the Appendix, where one can find the *best possible* way to *invert* into a $\mathcal{C}^\infty$-smooth function any given $\mathcal{C}^\infty$-smooth surjective *local* diffeomorphism which, in general, may fail to be injective.
This best way, however, does lead to a whole *family* of local inverses, even in the case of the inversion of one single such $\mathcal{C}^\infty$-smooth surjective local diffeomorphism. The effect is that, as a result of a Lie group action, we may again get a family of $\mathcal{C}^\infty$-smooth functions, and not only one single such function, as it happens in the case of projectable Lie group actions, see (2.3) - (2.7).
However, as mentioned, this proves to be unavoidable already in the case of a nonprojectable Lie group action on one single function, if - unlike it is usually done , Olver [1,2] - we try to *collect* together all of the local actions which can be defined in a classical manner.

When in the sequel, we go from functions to generalised functions, two more levels of *complexity* will have to come in. Namely, we shall have to deal with *sequences* of functions, instead of single functions. Further, we shall have to make the respective constructions for sequences of functions hold as well for *equivalence classes* of such sequences of functions, since such equivalence clasess are precisely the generalised functions of interest to us.

As indicated at the end of chapter 3, we shall consistently pursue the *two track* approach, where we present the theory for the usual nonparametric approach, and then also, for the approach based on the parametric representation.
It should be mentioned that Theorem 1 at the end of chapter 5, and its extension to generalised functions, given in chapter 9, will play an important role in mediating the natural link between the sides of the mentioned two track approach.
This chapter, however, is mostly dedicated to the first track approach.

For that purpose, however, let us recall the necessary basic facts about the nowhere dense differential algebras A in (1.6). More details are presented in Rosinger [6-9].

1. Nowhere Dense Algebras

As mentioned, one of the essential properties of the nowhere dense differential algebras of generalised functions in (1.6) is that they can treat in a 'smooth' manner singularities concentrated on arbitrary closed and nowhere dense subsets, subsets which, as is known, can have arbitrary large positive Lebesgue measure.

Given a nonvoid open $\Omega \subseteq \mathbf{R}^n$, the *nowhere dense ideal* $\mathcal{I}_{nd}(\Omega)$ in (1.6) is the crucial element of the construction, which will allow us to *disegard* the mentioned kind of singularities, see further details in chapter 12. This ideal is defined as the set of all the sequences

$$w = (w_\nu | \nu \in \mathbf{N})$$

of functions $w_\nu \in \mathcal{C}^\infty(\Omega) = \mathcal{C}^\infty(\Omega, \mathbf{R})$, with $\nu \in \mathbf{N}$, which satisfy the following *asymptotically dense vanishing* type property

$$(7.1.1) \qquad \begin{array}{l} \exists\ \Gamma \subset \Omega, \ \text{ closed, nowhere dense :} \\ \forall\ x \in \Omega \backslash \Gamma : \\ \exists\ \Delta \subseteq \Omega \backslash \Gamma \ \text{ neighbourhood of } x, \ \mu \in \mathbf{N} : \\ \forall\ \nu \in \mathbf{N}, \nu \geq \mu : \\ \qquad w_\nu = 0 \ \text{ on } \ \Delta \end{array}$$

It is easy to see that indeed $\mathcal{I}_{nd}(\Omega)$ is an ideal in the associative and commutative algebra $(\mathcal{C}^\infty(\Omega))^{\mathbf{N}}$, when considered with the respective termwise operations on sequences of functions.

As the second, and last, step in constructing the differential algebras A in (1.6), we take *any* subalgebra $\mathcal{A}_\Omega$ in $(\mathcal{C}^\infty(\Omega))^{\mathbf{N}}$, such that

$$(7.1.2) \qquad \mathcal{I}_{nd}(\Omega) \cup \mathcal{U}(\Omega) \subseteq \mathcal{A}_\Omega$$

where $\mathcal{U}(\Omega)$ is the set of all the sequences of functions

$$u(\psi) = (\psi, \psi, \psi, \ldots, \psi, \ldots)$$

with identical terms $\psi \in \mathcal{C}^\infty(\Omega)$. In other words, $\mathcal{U}(\Omega)$ is the *diagonal* of the infinite Cartesian product $(\mathcal{C}^\infty(\Omega))^{\mathbf{N}}$.
Obviously, $\mathcal{U}(\Omega)$ is a subalgebra of $(\mathcal{C}^\infty(\Omega))^{\mathbf{N}}$, and as such, it is isomorphic with $\mathcal{C}^\infty(\Omega)$, through the mapping $\psi \longmapsto u(\psi)$.
Clearly, the condition (7.1.2) can easily be satisfied by taking, for instance

$$(7.1.3) \qquad \mathcal{A}_\Omega = (\mathcal{C}^\infty(\Omega))^{\mathbf{N}}$$

In general, (7.1.1), (7.1.2) imply that the *quotient*

$$(7.1.4) \qquad A_\Omega = \mathcal{A}_\Omega \ / \mathcal{I}_{nd}(\Omega)$$

is an associative and commutative algebra with the *unit* element

$$(7.1.5) \qquad 1_{A_\Omega} = u(1) + \mathcal{I}_{nd}(\Omega) \in A_\Omega$$

We also obtain the algebra *embedding*

$$(7.1.6) \qquad \mathcal{C}^\infty(\Omega) \ni \psi \longmapsto u(\psi) + \mathcal{I}_{nd}(\Omega) \in A_\Omega$$

as a consequence of the following *neutrix* relation, see Rosinger [6-9], with which it is obviously equivalent

$$(7.1.7) \quad \mathcal{I}_{nd}(\Omega) \cap \mathcal{U}(\Omega) = \{0\}$$

and which also means that, in a geometric sense, the nowhere dense ideal $\mathcal{I}_{nd}(\Omega)$ is *off diagonal* in $(\mathcal{C}^\infty(\Omega))^{\mathbf{N}}$.

For convenience, in the sequel, we shall assume that the particular situation in (7.1.3) holds. However, it can be mentioned that, at the price of rather simple and obvious technical modifications, all the constructions in the sequel can be extended to the general case of (7.1.2).

It is easy to see that, for $p \in \mathbf{N}^n$, the mappings

$$(7.1.8) \quad \begin{array}{c} A_\Omega \xrightarrow{D^p} A_\Omega \\ s + \mathcal{I}_{nd}(\Omega) \longmapsto D^p s + \mathcal{I}_{nd}(\Omega) \end{array}$$

where for $s = (\psi_\nu | \nu \in \mathbf{N}) \in \mathcal{A}_\Omega = (\mathcal{C}^\infty(\Omega))^{\mathbf{N}}$ we have defined

$$(7.1.9) \quad D^p s = (D^p \psi_\nu | \nu \in \mathbf{N}) \ , \quad p \in \mathbf{N}^n$$

will turn A_Ω into a *differential* algebra. Moreover, the algebra embedding (7.1.6) will preserve the usual differentiation of $\mathcal{C}^\infty$-smooth functions.
Finally, we shall have the vector space embedding, see (1.7)

$$(7.1.10) \quad \mathcal{D}'(\Omega) \subset A_\Omega$$

which preserves the algebraic and differential operations on $\mathcal{C}^\infty$-smooth functions.

Our aim in this chapter is to *extend* the arbitrary nonlinear *parametric* Lie group actions (4.1) - (4.4) on functions, to actions

$$(7.1.11) \quad G \times A_\Omega \longrightarrow A_\Omega$$

that is, to define such arbitrary Lie group actions on generalised functions in the nowhere dense differential algebras A_Ω.

Here, the quotient structure (7.1.4) of the algebras A_Ω, and above all, the special structure of the ideals $\mathcal{I}_{nd}(\Omega)$, will allow us to obtain the extensions (7.1.11) in a natural and easy manner.

2. Lie Group Actions on Sequences of Functions

As a first step, we shall *lift* termwise the arbitrary Lie group actions (4.1) which act on individual functions, to actions on sequences of functions. Here, however, it will be useful to do so first on suitable subsets of functions in $\mathcal{C}_n^\infty(M)$. To that end, we shall start by simply lifting the parametric representation of functions given in (3.8), namely

$$(7.2.1) \quad \mathcal{C}^\infty(\Omega) \ni U \longmapsto U_* \in \mathcal{C}^\infty(\Omega, M)$$

to an embedding of sequences of functions

$$(7.2.2) \quad (\mathcal{C}^\infty(\Omega))^{\mathbf{N}} \ni s \longmapsto s_* \in (\mathcal{C}^\infty(\Omega, M))^{\mathbf{N}}$$

where for $s = (\psi_\nu | \nu \in \mathbf{N})$, with $\psi_\nu \in \mathcal{C}^\infty(\Omega)$, for $\nu \in \mathbf{N}$, we have set

$$(7.2.3) \quad s_* = ((\psi_\nu)_* | \nu \in \mathbf{N})$$

The important point to note about (7.2.2), (7.2.3) is that all of the terms $(\psi_\nu)_*$, with $\nu \in \mathbf{N}$, have the same domain of definition and the same range, see (3.5), namely

$$(7.2.4) \quad \Omega \overset{(\psi_\nu)_*}{\longrightarrow} M, \text{ with } \nu \in \mathbf{N}$$

Here, for further reference, we recall that we have the embeddings, see (3.6), (3.11)

$$(7.2.5) \quad \mathcal{C}^\infty(\Omega, M) \subseteq \mathcal{C}_{par}^\infty(\Omega, M) \subseteq \mathcal{C}_n^\infty(M)$$

owing to the fact that $\Omega \subseteq \mathbf{R}^n$ is nonvoid, open.

Now, since $M = \Omega \times \mathbf{R} \subseteq \mathbf{R}^{n+1}$, it follows that $\mathcal{C}^\infty(\Omega, M)$ is only a vector space, and *not* an algebra, when considered with the usual pointwise operations on functions. There is, however, a simple and natural way in which suitable subsets of $\mathcal{C}^\infty(\Omega, M)$, and in fact of $\mathcal{C}_{par}^\infty(\Omega, M)$ or $\mathcal{C}_n^\infty(M)$, can be endowed with a structure of algebra. Indeed, we note that the embedding (7.2.1) is *compatible* with the algebra operations. Namely, given $U, V \in \mathcal{C}^\infty(\Omega), \lambda \in \mathbf{R}$, we have

$$(7.2.6) \quad (\lambda U)_* = \lambda(U_*), \qquad (U+V)_* = U_* + V_*$$

as well as

$$(7.2.7) \quad (UV)_* = (U_*)(V_*)$$

where the product in the right hand term is well defined, owing to the fact that both U_* and V_* are parametrised in the *same* way, namely, see (3.5).

$$(7.2.8) \quad U_*(x) = (x, U(x)), \qquad V_*(x) = (x, V(x)), \ x \in \Omega$$

while on the other hand $U(x), V(x) \in \mathbf{R}$, for x $\in \Omega$, therefore, we can define

$$(7.2.9) \quad ((U_*)(V_*))(x) = (x, U(x)V(x)), \ x \in \Omega$$

and in this way (7.2.7) follows.

Let us for the moment look at this phenomenon in a more general manner. Given any $\Lambda \subseteq \mathbf{R}^n$ nonvoid, open, and a $\mathcal{C}^\infty$-smooth function $\Lambda \xrightarrow{P} \Omega$, we define

$$(7.2.10) \quad \mathcal{C}^\infty_P(\Lambda, M)$$

as the set of all $Q \in \mathcal{C}^\infty(\Lambda, M)$, for which there exists $q \in \mathcal{C}^\infty(\Lambda)$, such that

$$(7.2.11) \quad Q(y) = (P(y), q(y)), \qquad y \in \Lambda$$

We then say that Q in (7.2.11) has the *parametrisation* P. Now similar with (7.2.6) - (7.2.9), one can show that

$$(7.2.12) \quad \mathcal{C}^\infty_P(\Lambda, M) \text{ is an associative, commutative algebra}$$

Indeed, given $Q, Z \in \mathcal{C}^\infty_P(\Lambda, M)$ and $\lambda \in \mathbf{R}$, then for suitable $q, z \in \mathcal{C}^\infty(\Lambda, \mathbf{R})$, we have for $y \in \Lambda$, see (7.2.11)

$$(7.2.13) \quad Q(y) = (P(y), q(y)), \qquad Z(y) = (P(y), z(y))$$

and therefore, we can define for $y \in \Lambda$

$$(7.2.14) \quad \begin{array}{ll} (\lambda Q)(y) & = (P(y), \lambda q(y)) \\ (Q + Z)(y) & = (P(y), q(y) + z(y)) \\ (QZ)(y) & = (P(y), q(y) z(y)) \end{array}$$

Clearly, (7.2.6) - (7.2.9) corresponds to the particular case when in (7.2.12), we have

$$(7.2.15) \quad \Lambda = \Omega, \quad P = id_\Omega$$

since in view of (3.5), if $U \in \mathcal{C}^\infty(\Omega, \mathbf{R})$, then $U_* \in \mathcal{C}^\infty(\Omega, M)$, and the parametrisation of U_* is given by id_Ω. Moreover, the mapping

$$(7.2.16) \quad \mathcal{C}^\infty(\Omega) \ni U \longmapsto U_* \in \mathcal{C}^\infty_{id_\Omega}(\Omega, M)$$

is an *algebra isomorphism*, this being a stronger form of (7.2.1). Thus, we can return to (7.2.2) and write it as well in the *stronger* form: the mapping

$$(7.2.17) \quad (\mathcal{C}^\infty(\Omega))^{\mathbf{N}} \ni s \longmapsto s_* \in (\mathcal{C}^\infty_{id_\Omega}(\Omega, M))^{\mathbf{N}}$$

is an *algebra isomorphism*.

Now, coming back to the extension of the Lie group actions to sequences of functions, we note that in view of (4.1), (4.2) and (7.2.17), for given $g \in G$, we can define, see (7.2.3)

$$(7.2.18) \quad (\mathcal{C}^\infty(\Omega))^{\mathbf{N}} \ni s \quad \longmapsto \quad g\, s_* \in (\mathcal{C}^\infty(\Omega, M))^{\mathbf{N}}$$

where

$$gs_* \; = \; (\; g(\psi_\nu)_* \mid \nu \in \mathbf{N} \;)$$

that is, we obtain the Lie group action gs_*, by *lifting* termwise the arbitrary nonlinear parametric Lie group action (4.1) on functions, to the corresponding action on sequences of functions.

However, it is important to note that (4.1) - (4.4) allow for a *general* pattern of Lie group actions on sequences of functions, a pattern which underlies (7.2.18). Indeed, we can obviously define the Lie group action

(7.2.19) $\quad G \times (\mathcal{C}_n^\infty(M))^{\mathbf{N}} \longrightarrow (\mathcal{C}_n^\infty(M))^{\mathbf{N}}$

where for $g \in G$ and $V = (V_\nu | \nu \in \mathbf{N}) \in (\mathcal{C}_n^\infty(M))^{\mathbf{N}}$, we set

(7.2.20) $\quad gV = (gV_\nu | \nu \in \mathbf{N})$

And in view of (7.2.5), we obtain the following particular cases of action

$$\begin{array}{lccc} & G \times (\mathcal{C}^\infty(\Omega, M))^{\mathbf{N}} & \longrightarrow & (\mathcal{C}^\infty(\Omega, M))^{\mathbf{N}} \\ (7.2.21) & & & \\ & G \times (\mathcal{C}_{par}^\infty(\Omega, M))^{\mathbf{N}} & \longrightarrow & (\mathcal{C}_{par}^\infty(\Omega, M))^{\mathbf{N}} \end{array}$$

However, here the following *difficulty* arises: even in the particular case of the functions in the sequence $g \; s_*$ in (7.2.18), they do *no longer* have the same parametrisation with one another. Indeed, in view of (7.2.3), (7.2.20), it follows that

$$g \; s_* = (g(\psi_\nu)_* | \nu \in \mathbf{N})$$

and then (1.2) yields for $\nu \in \mathbf{N}$

$$(g(\psi_\nu)_*)(x) = (g_1(x, \psi_\nu(x)), g_2(x, \psi_\nu(x))), \;\; x \in \Omega$$

which means that the parametrisation, see (7.2.10), (7.2.11)

$$\Omega \ni x \longmapsto g_1(x, \psi_\nu(x)) \in \Omega$$

of $g(\psi_\nu)_*$ will depend on ψ_ν, when the Lie group action is nonprojectable. In this way, the algebra properties (7.2.6) - (7.2.17) can no longer be secured, once the general group action (7.2.19), (7.2.20), or its particular cases (7.2.18), (7.2.21), are performed.

3. Lie Group Actions on Generalised Functions, Preliminaries

The above mentioned loss, however, proves to be unimportant. Indeed, it turns out that the essential feature we need is related to the somewhat extended concept of *equivalence relation* $\approx_{nd}$ defined next, which is suggested by the more general - rather asymtotic or topological, than algebraic type - structure of the nowhere dense ideal $\mathcal{I}_{nd}(\Omega)$,.

The natural domain of definition of this extended concept of equivalence relation $\approx_{nd}$ will therefore not be $(\mathcal{C}^\infty(\Omega))^{\mathbf{N}}$, as may at first be suggested by (7.1.1) - (7.1.4) or (7.2.2). Rather, in view of chapters 4 and 5, we shall define the *symmetric and transitive*, but not necessarily reflexive binary relation $\approx_{nd}$ on $(\mathcal{C}^\infty_{par}(\Omega, \mathbf{R}^l))^{\mathbf{N}}$, with $l \in \mathbf{N}, l \geq 1$, as follows. Given

$$s = (\psi_\nu | \nu \in \mathbf{N}),\ t = (\chi_\nu | \nu \in \mathbf{N}) \in (\mathcal{C}^\infty_{par}(\Omega, \mathbf{R}^\ell))^{\mathbf{N}}$$

then we set

$$(7.3.1) \quad s \approx_{nd} t$$

if and only if there exists $\Delta \subseteq \Omega$ nonvoid, open, such that

$$(7.3.2) \quad \psi_\nu,\ \chi_\nu\ :\ \Delta \longrightarrow \mathbf{R}^\ell\ ,\ \ \nu \in \mathbf{N}$$

and also, see (7.1.1)

$$(7.3.3) \quad \begin{array}{l} \exists\ \Gamma\ \subset\ \Delta,\ \text{ closed, nowhere dense }\ : \\ \forall\ x\ \in\ \Delta \setminus \Gamma\ : \\ \exists\ \Delta'\ \subseteq\ \Delta \setminus \Gamma\ \text{ neighbourhood of } x\ ,\ \mu\ \in\ \mathbf{N}\ : \\ \forall\ \nu\ \in\ \mathbf{N},\ \nu \geq \mu\ : \\ \\ \qquad \psi_\nu = \chi_\nu \ \text{ on }\ \Delta' \end{array}$$

We note that indeed $\approx_{nd}$ is not reflexive on $(\mathcal{C}^\infty_{par}(\Omega, \mathbf{R}^l))^{\mathbf{N}}$.
However, in order to avoid unnecessary complication in notation, it will be useful to introduce here a *slight extension* of the usual concepts of equivalence relation and quotient space, see (7.3.18) - (7.3.23) at the end of this section. In this way, we shall be able to keep the notation simple, namely, the same with the usual one. Also, in the sense specified at the end of this section, it will still allow us to call $\approx_{nd}$ an equivalence relation.

In the sense of the above, it will be useful, for given $s \in (\mathcal{C}^\infty_{par}(\Omega, \mathbf{R}^\ell))^{\mathbf{N}}$, to denote by, see (7.3.23)

$$(7.3.4) \quad \text{class}_{nd}\ s$$

the $\approx_{nd}$ equivalence class of s.

A first obvious and useful property of the equivalence relation $\approx_{nd}$ is as follows, see (7.2.2), (7.2.17)

$$(7.3.5) \quad \begin{array}{l} \forall\ s, t \in (\mathcal{C}^\infty(\Omega))^{\mathbf{N}} \ : \\ s \approx_{nd} t \Longrightarrow s_* \approx_{nd} t_* \end{array}$$

As a consequence of (7.3.5), we can define the *embedding*

$$(7.3.6) \quad A_\Omega \ni U \longmapsto U_* \in (\mathcal{C}^\infty(\Omega, M))^{\mathbf{N}} / \approx_{nd}$$

as follows : if

$$(7.3.7) \quad U = s + \mathcal{I}_{nd}(\Omega)$$

with $s \in (\mathcal{C}^\infty(\Omega))^{\mathbf{N}}$, then

$$(7.3.8) \quad U_* = \text{class}_{nd}\ s_*$$

The second property of interest of the equivalence relation $\approx_{nd}$ is that it allows us to define the arbitrary Lie group actions

$$(7.3.9) \quad G \times ((\mathcal{C}^\infty(\Omega, M))^{\mathbf{N}} / \approx_{nd}) \longrightarrow (\mathcal{C}^\infty(\Omega, M))^{\mathbf{N}} / \approx_{nd}$$

as follows : given $g \in G$ and

$$(7.3.10) \quad V = \text{class}_{nd}(V_\nu | \nu \in \mathbf{N}) \in (\mathcal{C}^\infty(\Omega, M))^{\mathbf{N}} / \approx_{nd}$$

where $V_\nu \in \mathcal{C}^\infty(\Omega, M)$, for $\nu \in \mathbf{N}$, then, see (4.1), (7.2.19), (7.2.20), we shall define

$$(7.3.11) \quad g\ V = \text{class}_{nd}(g\ V_\nu | \nu \in \mathbf{N})$$

and in view of (4.1), (7.3.5), it is easy to see that this definition is indeed correct.

Finally, we are in the position to try to define the arbitrary nonlinear Lie group actions

$$(7.3.12) \quad G \times ((\mathcal{C}^\infty_{par}(\Omega))^{\mathbf{N}} / \approx_{nd}) \xrightarrow{\ \mathcal{L}\ } (\mathcal{C}^\infty_{par}(\Omega))^{\mathbf{N}} / \approx_{nd}$$

as the *composition* of the following *three* mappings : first the mapping

$$(7.3.13) \quad G \times ((\mathcal{C}^\infty_{par}(\Omega))^{\mathbf{N}}) / \approx_{nd}) \xrightarrow{\ id_G \times \beta_*\ } G \times ((\mathcal{C}^\infty_{par}(\Omega, M))^{\mathbf{N}} / \approx_{nd})$$

followed by the second mapping given by the obvious extension of (7.3.9), namely

$$(7.3.14) \quad G \times ((\mathcal{C}^\infty_{par}(\Omega, M))^{\mathbf{N}} / \approx_{nd}) \xrightarrow{\ \mathcal{L}_{prm}\ } (\mathcal{C}^\infty_{par}(\Omega, M))^{\mathbf{N}} / \approx_{nd}$$

and finally, followed by the third mapping

$$(7.3.15) \quad (\mathcal{C}^\infty_{par}(\Omega, M))^{\mathbf{N}} / \approx_{nd} \xrightarrow{\ \delta\ } (\mathcal{C}^\infty_{par}(\Omega))^{\mathbf{N}} / \approx_{nd}$$

which will be defined rigorously in subsection 7.4.

Here we have to note that the *only difficulty* we shall encounter with the three mappings (7.3.13) - (7.3.15) is related to the third one, that is, δ in (7.3.15). And this difficulty is *not at all* connected with any kind of Lie group action. In fact, as seen in (7.3.14), based on (4.1), we can easily define such Lie group actions, and do so without any restrictions whatsoever.
On the other hand, the difficulty with the mapping δ in (7.3.15) is only related to the problem of *inverting* into a $\mathcal{C}^\infty$-smooth function a given arbitrary $\mathcal{C}^\infty$-smooth local diffeomorphism, which thus may fail to be injective. And as seen in the Appendix, this leads to certain unavoidable complications, even in the case of one single such $\mathcal{C}^\infty$-smooth local diffeomorphism, let alone when we have to deal with *equivalence classes of sequences* of $\mathcal{C}^\infty$-smooth local diffeomorphisms, which are supposed to be the elements in the domain of definition of δ.
Therefore, the representation of δ in (7.3.15) is only *tentative*, and it will have to be modified later in section 7.4, in order to make it precise. This will be done by having its domain of definition *restricted*, while at the same time, having its range *extended.*
Needless to say, this modification will accordingly affect the definition of the mapping (7.3.12), which in fact is the main object of our attention. See below Remark 2, for further comments.

In (7.3.13) we used the mapping

$$(7.3.16) \qquad (\mathcal{C}^\infty_{par}(\Omega))^{\mathbf{N}} / \approx_{nd} \xrightarrow{\beta_*} (\mathcal{C}^\infty_{par}(\Omega, M))^{\mathbf{N}} / \approx_{nd}$$

which is the following natural extension of (7.3.6) : we note that, see (7.1.3), (7.1.4), (7.3.1), (7.3.2), (7.3.3), and also (7.3.21)

$$(7.3.17) \qquad \begin{aligned} (\mathcal{C}^\infty_{par}(\Omega))^{\mathbf{N}} / \approx_{nd} \quad &= \quad \bigcup_{\phi \neq \Delta \subseteq \Omega,\ \Delta \text{ open}} A_\Delta \\ (\mathcal{C}^\infty_{par}(\Omega, M))^{\mathbf{N}} / \approx_{nd} \quad &= \quad \bigcup_{\phi \neq \Delta \subseteq \Omega,\ \Delta \text{ open}} ((\mathcal{C}^\infty(\Delta, M))^{\mathbf{N}} / \approx_{nd}) \end{aligned}$$

Hence β_* in (7.3.16) is simply the extension of the mapping in (7.3.6), where Ω is replaced with any $\Delta \subseteq \Omega$, nonvoid, open, and the unions in the right hand terms of (7.3.17) are each considered to be unions of pairwise disjoint sets.

Remark 2

We aim to attain the main goal of this chapter, namely, to define the arbitrary nonlinear parametric Lie group actions (7.3.12), see also its particular case in (7.1.11), through the composition of the three mappings (7.3.13) - (7.3.15). In view of (7.3.17) and (7.1.11), the mapping (7.3.12) aims, among others, to define arbitrary nonlinear Lie group actions on the nowhere dense differential algebras (7.1.4). Now, among the three mappings (7.3.13) - (7.3.15), only the mapping (7.3.14) does involve Lie group actions. However, owing to the extreme *simplicity* of the way arbitrary nonlinear Lie group actions were defined in (4.1) - (4.4) on *parametric* functions, the definition of the mapping (7.3.14) is a rather trivial matter, although it does not require any restriction on the Lie groups involved. The mapping (7.3.13) obviously reduces to (7.3.16), therefore, in essence, it is but about the very simple mapping (3.8), which associates to each function

$$\Delta \xrightarrow{U} \mathbf{R}$$

its *parametric* representation

$$\Delta \xrightarrow{U_*} M$$

where $\Delta \subseteq \Omega$ is nonvoid, open, and $M = \Omega \times \mathbf{R}$. The critical point of the attempted construction in (7.3.12) - (7.3.15) is the mapping δ in (7.3.15). However, δ does *not* involve any kind of Lie group action, and it only aims to revert, or *invert*, the earlier action of β_* in (7.3.13) or (7.3.16). In this way, δ is in its essence but about *extracting all* possible $\mathcal{C}^\infty$-smooth functions

$$\Delta' \xrightarrow{U} \mathbf{R}$$

from a given $\mathcal{C}^\infty$-smooth *parametric* function

$$\Delta \xrightarrow{V} M$$

with $\Delta,\ \Delta' \subseteq \Omega$ nonvoid, open, and M as above. Of course, the difficulty here is coming from the fact that δ has to operate *simultaneously* on the terms of sequences of parametric functions, and do so in a manner *compatible* with the quotient structure defined by the equivalence relation $\approx_{nd}$. Based on the Appendix, see there (A.20), as well as (A.13) - (A.19), (A.21), and also (A.30), (A.36), (A.39), one is given the general way to do this extraction in the case of a *single* parametric function. Here in this chapter, therefore, our task is to construct the mapping δ by *lifting* this earlier result from the case of a single function, to that of the *quotient structures* involved in (7.3.15).

□

Let us present here the mentioned *slight extension* of the concepts of equivalence relation and quotient space. Let X be an arbitrary nonvoid set, while $r \subseteq X \times X$ is a *symmetric and transitive* binary relation on X, which need *not* be reflexive as well. For elements $x, y \in X$, we shall as customary write $(x, y) \in r$, or equivalently, $x\ r\ y$.
Let us now denote

$$(7.3.18) \quad X_0 \;=\; \{\, x \in X \mid x\ r\ x \,\}$$

that is, X_0 is the set of elements $x \in X$ which happen to be *reflexive* with respect to the given binary relation r. We shall assume that $X_0 \neq \phi$.
Further, we denote by r_0 the restriction of r to X_0, that is, $r_0 \;=\; r \cap (X_0 \times X_0)$.

Then clearly, r_0 is a usual *equivalence* relation on X_0.

In this way, X_0/r_0 is a usual quotient space, and for the convenience of simplicity, we shall denote

$$(7.3.19) \quad X/r = X_0/r_0$$

In our case in (7.3.1) - (7.3.3), we obviously have

$$(7.3.20) \quad \begin{aligned} X &= (\mathcal{C}^\infty_{par}(\Omega, \mathbf{R}^l))^{\mathbf{N}} \\ r &= \approx_{nd} \\ X_0 &= \bigcup_{\phi \neq \Delta \subseteq \Omega,\ \Delta \text{ open}} (\mathcal{C}^\infty(\Delta, \mathbf{R}^l))^{\mathbf{N}} \end{aligned}$$

and therefore (7.3.19) becomes

$$(7.3.21) \quad (\mathcal{C}^\infty_{par}(\Omega, \mathbf{R}^l))^{\mathbf{N}} / \approx_{nd} = \left(\bigcup_{\phi \neq \Delta \subseteq \Omega,\ \Delta \text{ open}} (\mathcal{C}^\infty(\Delta, \mathbf{R}^l))^{\mathbf{N}} \right) / \approx_{nd}$$

where in the right hand term, we have used the usual concept of equivalence relation and quotient space.
We note that (7.3.21) can be written as well in the form

$$(7.3.22) \quad (\mathcal{C}^\infty_{par}(\Omega, \mathbf{R}^l))^{\mathbf{N}} / \approx_{nd} = \bigcup_{\phi \neq \Delta \subseteq \Omega,\ \Delta \text{ open}} ((\mathcal{C}^\infty(\Delta, \mathbf{R}^l))^{\mathbf{N}} / \approx_{nd})$$

with the same usual concept of equivalence relation and quotient space.

Finally, for $x \in X$, we shall denote by

$$(7.3.23) \quad \text{class}_r\, x = \{\, y \in X \mid y\, r_0\, x \,\}$$

the set of elements $y \in X$ which are in the relation r_0 with x. Clearly, when $x \in X_0$, then $\text{class}_r\, x$ reduces to the usual r_0 equivalence class of x in X_0. Otherwise, simply, $\text{class}_r\, x = \phi$.

Now, in the slightly extended sense specified in (7.3.18) - (7.3.23), we shall for the convenience of simplicity still call r an *equivalence relation*, X/r its *quotient space*, and for $x \in X$, $\text{class}_r\, x$ the *equivalence class* of x in X.

4. Lie group Actions on Generalised Functions, Definition

We return now to the problem of the precise definition of the mapping δ in (7.3.15), and note that in view of (7.3.17), this mapping tentatively becomes

$$(7.4.1) \quad \bigcup_{\phi \neq \Delta \subseteq \Omega,\ \Delta \text{ open}} ((\mathcal{C}^\infty(\Delta, M))^{\mathbf{N}} / \approx_{nd}) \quad \xrightarrow{\delta} \quad \bigcup_{\phi \neq \Delta \subseteq \Omega,\ \Delta \text{ open}} A_\Delta$$

4.1. Family of Commutative Diagrams Associated to a Function

Let us first adapt and somewhat extend a few of the notations used in the previous sections, or in the Appendix. Given

(7.4.2) $\Lambda \xrightarrow{V} M$

an element in $\mathcal{C}_n^\infty(M)$, it is clear that it equivalently gives the mappings, see (3.19)

$$\begin{array}{ll} & \Lambda \ni y \longmapsto V(y) = (V_1(y), V_2(y)) \in M \\ (7.4.3) & \Lambda \ni y \longmapsto V_1(y) \in \Omega \\ & \Lambda \ni y \longmapsto V_2(y) \in \mathbf{R} \end{array}$$

which belong to $\mathcal{C}^\infty(\Lambda, M), \mathcal{C}^\infty(\Lambda, \Omega)$ and $\mathcal{C}^\infty(\Lambda) = \mathcal{C}^\infty(\Lambda, \mathbf{R})$, respectively. In case we want to turn the parametric function (7.4.2) into an equivalent usual function

(7.4.4) $\Delta' \xrightarrow{U} \mathbf{R}$

in $\mathcal{C}_{par}^\infty(\Omega)$, that is, with $\Delta' \subseteq \Omega$ nonvoid, open, then we have to consider the system of equations

$$(7.4.5) \quad \begin{array}{l} x = V_1(y) \\ U = V_2(y) \end{array}$$

where $y \in \Lambda$, $x \in \Omega$, $U \in \mathbf{R}$ and solve the first one in y, in order to obtain a nonvoid open set Δ', with

(7.4.6) $\Delta' = V_1(\Lambda) \subseteq \Omega$

on which we want to have the $\mathcal{C}^\infty$-smooth inverse

(7.4.7) $y = V_1^{-1}(x), \quad x \in \Delta'$

Once that is done, we substitute this y in the second equation (7.4.5), and finally get

(7.4.8) $U(x) = V_2(V_1^{-1}(x)), \quad x \in \Delta'$

However, in general, (7.4.7), (7.4.8) will only hold locally, since in (7.4.6) we shall not have $\Delta' = V_1(\Lambda)$, see (5.31) - (5.37).

It follows that, given (7.4.2), we are interested in the Jacobian of V_1 in (7.4.3), which as usual, we shall denote by $J_{V_1} : \Lambda \longrightarrow \mathbf{R}$. Then we consider the open set, see (5.33)

(7.4.9) $D_V = \{\, y \in \Lambda \mid J_{V_1}(y) \neq 0\}$

and in case $D_V \neq \phi$, see (5.26) - (5.30), and the comment following (5.75), we obtain the $\mathcal{C}^\infty$-smooth surjective local diffeomorphism

(7.4.10) $\quad D_V \xrightarrow{V_1} V_1(D_V) \subseteq \Omega$

where $V_1(D_V)$ will also be open. Further, we denote, see (A.5)

(7.4.11) $\mathcal{D}_V = \left\{ \Lambda' \subseteq D_V \;\middle|\; \begin{array}{l} 1) \;\; \Lambda' \text{ nonvoid, open} \\ 2) \;\; \Lambda' \xrightarrow{V_1} V_1(\Lambda') \\ \quad\;\; \text{is a } \mathcal{C}^\infty\text{-smooth diffeomorphism} \end{array} \right\}$

Now, in view of (7.4.4) - (7.4.8), we would ideally be interested in constructing a commutative diagram, see (A.16)

(7.4.12)
$$\begin{array}{ccc} \Delta' & \xrightarrow{\quad U \quad} & \mathbf{R} \\ & \nwarrow^{V_1} \quad {}^{V_2}\nearrow & \\ & \Lambda & \end{array}$$

Although in view of the Appendix, such a diagram (7.4.12) cannot in general be obtained for an arbitrary parametrically given function V in (7.4.2), the next best thing we can do is to construct for each such function V a whole *family* of commutative diagrams, see (A.17)

(7.4.13)
$$\begin{array}{ccc} \Delta' & \xrightarrow{\quad U \quad} & \mathbf{R} \\ & \nwarrow^{V_1} \quad {}^{V_2}\nearrow & \\ & \Lambda' & \end{array}$$

where $\Lambda' \in \mathcal{D}_V$ and, see (A.18)

(7.4.14) $\quad \Delta' = V_1(\Lambda'), \quad U = V_2 \circ V_1^{-1}$

with Λ' and Δ' nonvoid, open.

Here again, we have to recall that in (7.4.4) and (7.4.8), we are only interested in the respective nonvoid, open domains $\Delta' \subseteq \Omega$, and not in any other arbitrary domains of parameters, such as for instance Λ'.

Nevertheless, the family (7.4.13), (7.4.14) cannot in general be indexed by the nonvoid, open subsets $\Delta' \subseteq \Omega$, but only by $\Lambda' \in \mathcal{D}_V$, see the respective comments in the Appendix, which follow (A.19). This is then the *price* we have to pay for using the general parametric representations (7.4.2). And as it turns out, this is in fact the *only* price.

4.2. Uniform Family of Commutative Diagrams Associated to a Sequence of Functions

Let us now take an element in the tentative domain of δ in (7.4.1), that is

(7.4.15) $\quad V = \text{class}_{nd}(V_\nu | \nu \in \mathbf{N})$

where for a suitable nonvoid, open $\Delta \subseteq \Omega$, we have $V_\nu \in \mathcal{C}^\infty(\Delta, M)$, with $\nu \in \mathbf{N}$, in other words

(7.4.16) $\quad (V_\nu | \nu \in \mathbf{N}) \in (\mathcal{C}^\infty(\Delta, M))^{\mathbf{N}}$

Using the notation in (7.4.2), (7.4.3), we have for $\nu \in \mathbf{N}$

(7.4.17) $\quad \Delta \xrightarrow{V_\nu} M, \quad \Delta \xrightarrow{V_{\nu 1}} \Omega, \quad \Delta \xrightarrow{V_{\nu 2}} \mathbf{R}$

while (7.4.9) gives

(7.4.18) $\quad D_\nu = D_{V_\nu} = \{x \in \Delta \mid J_{V_{\nu 1}}(x) \neq 0\}$

with the mapping

(7.4.19) $\quad D_\nu \xrightarrow{V_{\nu 1}} V_{\nu 1}(D_\nu) \subseteq \Omega$

being a $\mathcal{C}^\infty$-smooth surjective local diffeomorphism, and with

(7.4.20) $\quad D'_\nu = V_{\nu 1}(D_\nu)$ nonvoid, open

if we assume that, see (5.26) - (5.30)

(7.4.21) $\quad D_\nu \neq \phi$

Further, for $\nu \in \mathbf{N}$, we obtain through (7.4.11)

(7.4.22) $$\mathcal{D}_\nu = \mathcal{D}_{V_\nu} = \left\{ \Delta''_\nu \subseteq D_\nu \;\middle|\; \begin{array}{l} 1)\ \Delta''_\nu \text{ nonvoid, open} \\ 2)\ \Delta''_\nu \xrightarrow{V_{\nu 1}} V_{\nu 1}(\Delta''_\nu) \\ \quad \text{is a } \mathcal{C}^\infty\text{-smooth diffeomorphism} \end{array} \right\}$$

and finally, (7.4.13), (7.4.14) give the *family* of commutative diagrams

(7.4.23)

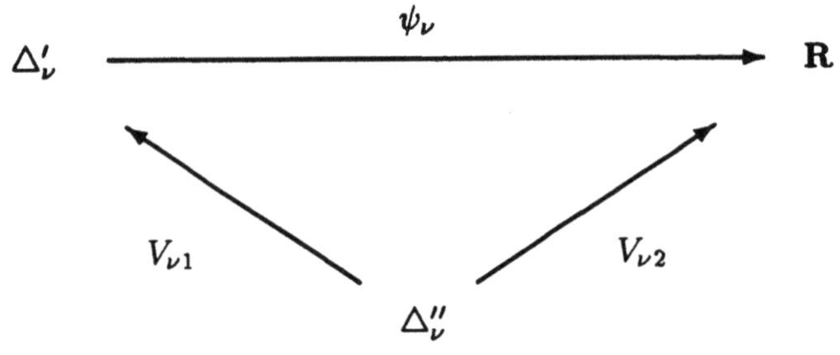

with $\Delta''_\nu \in \mathcal{D}_\nu$ and

(7.4.24) $\Delta'_\nu = V_{\nu 1}(\Delta''_\nu), \quad \psi_\nu = V_{\nu 2} \circ (V_{\nu 1} \mid_{\Delta''_\nu})^{-1}$

4.3. The Domain of Definition of δ

Having thus set up the necessary notations, let us now return to the problem of the precise definition of the mapping δ in (7.4.1). We shall start by specifying the precise domain of definition of δ, that is, by specifying certain *minimal* restrictions on its initially attempted domain in (7.4.1), so that its effective construction can indeed be carried out.
Let us, therefore, consider an arbitrary

(7.4.25) $V \in (\mathcal{C}^\infty(\Delta, M))^{\mathbf{N}} / \approx_{nd}$

in its tentative domain of definition, where $\Delta \subseteq \Omega$ is nonvoid, open. Our tentative aim in (7.4.1) is to obtain

(7.4.26) $\delta(V) \in A_{\Delta'}$

for a suitable nonvoid open $\Delta' \subseteq \Omega$. Hence, we are looking for a relation

(7.4.27) $\delta(V) = \mathrm{class}_{nd}(U'_\nu | \nu \in \mathbf{N})$

where

(7.4.28) $(U'_\nu | \nu \in \mathbf{N}) \in (\mathcal{C}^\infty(\Delta'))^{\mathbf{N}}$

As mentioned in Remark 2, the mapping δ is a sort of inverse of β_* in (7.3.16). This means that for a suitable representation

(7.4.29) $V = \mathrm{class}_{nd}(V_\nu | \nu \in \mathbf{N})$

where

(7.4.30) $(V_\nu | \nu \in \mathbf{N}) \in (\mathcal{C}^\infty(\Delta, M))^{\mathbf{N}}$

the mapping δ is supposed to act in such a way that the following *family* of commutative diagrams of $\mathcal{C}^\infty$-smooth functions hold for $\nu \in \mathbf{N}$

(7.4.31)

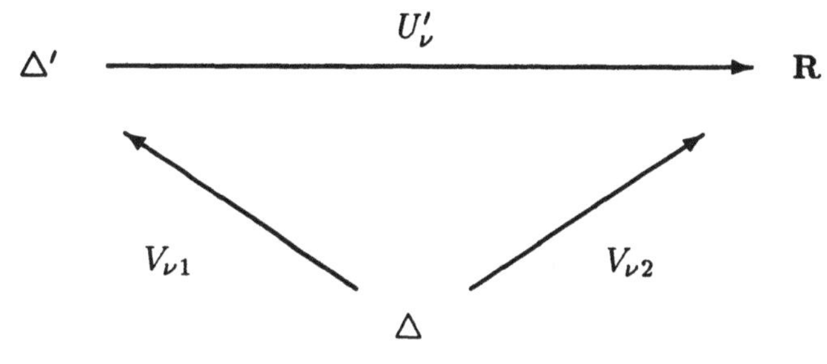

where $\Delta \in \mathcal{D}_\nu$.

These diagrams are supposed to be the *uniformised* versions of (7.4.23), with respect to $\nu \in \mathbf{N}$, as far as the two nonvoid and open domains Δ and Δ' are concerned.
However, as seen in (7.4.12) - (7.4.14), and correspondingly in (7.4.23), (7.4.24), diagrams of the form (7.4.31) *cannot* in general be obtained.

A way to relax these diagrams (7.4.31) in order to make them possible, is presented next, and it still amounts to a certain uniform version of (7.4.23). This way is quite clearly suggested by (7.4.26), (7.4.27), in the sense that what is really needed in (7.4.23) to be *independent* of $\nu \in \mathbf{N}$, are the subsets Δ'_ν, and not so much the subsets Δ''_ν. Namely, we shall consider the commutative diagrams of $\mathcal{C}^\infty$-smooth functions of the form

(7.4.32)

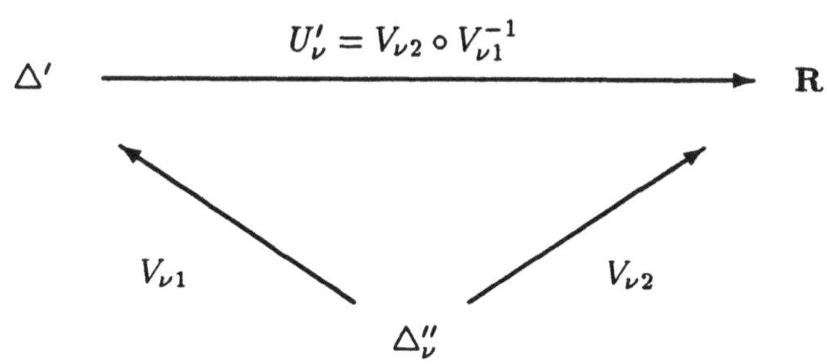

where, for $\nu \in \mathbf{N}$, we have in addition, see (7.4.14), (7.4.24)

$$\Delta' = V_{\nu 1}(\Delta''_\nu), \quad \Delta''_\nu \in \mathcal{D}_\nu = \mathcal{D}_{V_\nu} \tag{7.4.33}$$

Let us recapitulate. When we are given V in (7.4.25), and then we try to define $\delta(V)$ in (7.4.26), we are faced with the following three *multivalued* mappings

$$(7.4.34)\qquad \begin{array}{ll} & V \in (\mathcal{C}^\infty(\Delta, M))^{\mathbf{N}} / \approx_{nd} \\ 1 & \downarrow \\ & (V_\nu \mid \nu \in \mathbf{N}) \in (\mathcal{C}^\infty(\Delta, M))^{\mathbf{N}} \\ 2 & \downarrow \\ & \Delta' \subseteq \Omega \text{ in } (7.4.32),\ (7.4.33) \\ 3 & \downarrow \\ & U' = \text{class}_{nd}(U'_\nu \mid \nu \in \mathbf{N}) \in A_{\Delta'} \end{array}$$

Indeed, the multivalued mapping 1 appears naturally, owing to (7.4.29), that is, to the quotient structure of the nowhere dense algebras.

Further, the multivalued mapping 2 is already present in the case of one single function $V_\nu : \Delta \longrightarrow M$, with $\nu \in \mathbf{N}$ fixed, as seen in (7.4.12) - (7.4.14), (7.4.22) - (7.4.24), (7.4.31), (7.4.32), and this is due to the fact that the mapping (7.4.19) is in general *not* injective. Added to all this is the fact that we have to make the mapping 2 *uniform* with respect to $\nu \in \mathbf{N}$, see (7.4.32).

Finally, the multivalued mapping 3 appears owing to the fact that in (7.4.32), even in the case of one single function $V_\nu : \Delta \longrightarrow M$, with $\nu \in \mathbf{N}$ fixed, the resulting novoid open subsets $\Delta''_\nu \subseteq \Omega$, $\Delta''_\nu \in \mathcal{D}_\nu = \mathcal{D}_{V_\nu}$ are in general *not* determined uniquely by Δ'. This again is a consequence of the fact that the mapping (7.4.19) need not in general be injective, see for related details the earlier mentioned part of the Appendix.

The above leads naturally to the following *restriction* on the tentative domain of definition of δ in (7.4.1).

Given $\Delta \subseteq \Omega$ nonvoid, open, let us denote by

$$(7.4.35)\qquad B_\Delta$$

the set of all

$$(7.4.36)\qquad V \in (\mathcal{C}^\infty(\Delta, M))^{\mathbf{N}} / \approx_{nd}$$

to which one can associate

$$(7.4.37)\qquad \begin{array}{l} \Delta' \subseteq \Omega \text{ nonvoid, open} \\ U' \in A_{\Delta'} \end{array}$$

which have the property, see (7.4.34)

$$(7.4.38)\qquad \begin{array}{l} \forall\ x' \in \Delta' : \\ \exists\ V \stackrel{1}{\longmapsto} (V_\nu \mid \nu \in \mathbf{N}) \stackrel{2}{\longmapsto} \Delta^0 \stackrel{3}{\longmapsto} U^0 : \\ \quad *)\ x' \in \Delta^0 \subseteq \Delta',\ \Delta^0 \text{ open},\ U^0 \in A_{\Delta^0} \\ \quad **)\ U' = U^0 \text{ on } \Delta^0 \end{array}$$

Here we note that, in view of (7.1.1) - (7.1.10), given $U' \in A_{\Delta'}$ and $\Delta^0 \subseteq \Delta'$ nonvoid, open, the restriction of U' to Δ^0 can be defined in an obvious natural way.

Remark 3

1) Instead of condition (7.4.38) above, and as in fact may be suggested *directly* by (7.4.34), we could have used the simpler looking and apparently more natural condition

$$\begin{aligned} &\exists\ (V_\nu|\nu \in \mathbf{N}) \in (\mathcal{C}^\infty(\Delta, M))^{\mathbf{N}} \ : \\ &V \overset{1}{\longmapsto} (V_\nu|\nu \in \mathbf{N}) \overset{2}{\longmapsto} \Delta' \overset{3}{\longmapsto} U' \end{aligned} \tag{7.4.39}$$

However, it is obvious that

$$(7.4.39) \implies (7.4.38)$$

in the sense that, given

$$V \overset{1}{\longmapsto} (V_\nu|\nu \in \mathbf{N}) \overset{2}{\longmapsto} \Delta' \overset{3}{\longmapsto} U'$$

as in (7.4.39), then V, Δ', U' satisfy (7.4.38), since for $x' \in \Delta'$, we can all the time choose the same $(V_\nu|\nu \in \mathbf{N})$, while we take $\Delta^0 = \Delta'$, $U^0 = U'$.

Therefore, in the above definition of B_Δ, we prefer to use condition (7.4.38), since it is not weaker than (7.4.39). Thus, it may lead to a B_Δ which is not smaller. And clearly, we are interested in as large a B_Δ as possible, since that leads, see (7.4.40), (7.4.46) below, to a milder restriction on the domain of definition of δ.

Furthermore, (7.4.38) is obviously but a condition in which (7.4.39) is in fact present *locally*, that is, in a neighbourhood of each $x' \in \Delta'$. In this way, the nonvoid open subsets $\Delta' \subseteq \Delta$ which appear in (7.4.38), may be at least as *large* as those in (7.4.39).

However, the main reason we choose condition (7.4.38), and not (7.4.39), is that in the former, for given V, the dependence of Δ' and U' on the particular choice of

$$V \overset{1}{\longmapsto} (V_\nu|\nu \in \mathbf{N})$$

appears to be *less* strong, being only local.

2) It will be useful to recapitulate here certain *sufficient* steps, when for a given nonvoid open $\Delta \subseteq \Omega$ and

$$V \in (\mathcal{C}^\infty(\Delta, M))^{\mathbf{N}} / \approx_{nd}$$

we want to show that

$$V \in B_\Delta$$

so that we may compute, see (7.4.46) below

$$\delta(V)$$

According to (7.4.39) these steps are as follows. We first find

$$V \overset{1}{\longmapsto} (V_\nu | \nu \in \mathbf{N}) \in (\mathcal{C}^\infty(\Delta, M))^{\mathbf{N}}$$

such that, for each $\nu \in \mathbf{N}$, we then can find

$$\Delta''_\nu \in \mathcal{D}_\nu = \mathcal{D}_{V_\nu}$$

so that (7.4.32) and (7.4.33) hold for a suitable nonvoid open subset

$$\Delta' \subseteq \Omega$$

which does *not* depend on $\nu \in \mathbf{N}$. In this case, we can immediately obtain

$$U' = \text{class}_{nd}(U'_\nu | \nu \in \mathbf{N}) \in A_{\Delta'}$$

therefore, condition (7.4.39) will be satisfied.
Of course, we can use as well the weaker, and more involved, condition (7.4.38), which however is not only sufficient, but also necessary in view of the above given definition of B_Δ.

4.4. The Definition of δ

Finally, with the use of the building block B_Δ in (7.4.35), we can give the *precise domain of definition* of the δ, as follows

(7.4.40) $\qquad B_{[\Omega]} = \bigcup B_\Delta$

where the union is taken over all nonvoid and open $\Delta \subseteq \Omega$.

As far as the *precise range* of δ is concerned, we shall have to take into account the generally unavoidable *multivalued* nature of the mappings in (7.4.34), and thus as well in (7.4.38). This implies that the mapping

(7.4.41) $\qquad V \overset{4}{\longmapsto} (\Delta', U')$

given in (7.4.36), (7.4.37) is again *multivalued*. More precisely, the *single valued* version of (7.4.41) is

(7.4.42) $\quad V \overset{5}{\longmapsto} ((\Delta'_i, U'_i) | i \in I)$

where the index set I may depend on V, while at the same time we have $U'_i \in A_{\Delta'_i}$, for $i \in I$. Obviously, one can assume an upper limit on the cardinality of I, such as for instance

(7.4.43) $\quad \text{car}\, I \leq 2^{\text{car}\, \mathbf{R}}$

which results from the upper limit on the cardinality of the set of nonvoid open subsets Δ' in Ω, together with the upper limit on the cardinality of any of the nowhere dense differential algebras $A_{\Delta'}$. For what follows, however, it will be convenient to replace (7.4.43) with an equivalent condition

(7.4.44) $\quad I \subseteq J$

where J is a fixed set of an appropriate cardinality.

Therefore, once we take the domain of definition of δ as given in (7.4.40), the natural *range* of δ is given by the *families* of generalised functions in the nowhere dense differential algebras (7.1.4), which are elements of

(7.4.45) $\quad A^{[\Omega]} = \bigcup \prod A_{\Delta'_i}$

where the union is taken over all $I \subseteq J$, while for each given I, the product is taken over all $i \in I$. In this way, we can at last arrive at the *precise definition* of δ as being the mapping

(7.4.46) $$\begin{array}{ccc} B_{[\Omega]} & \xrightarrow{\delta} & A^{[\Omega]} \\ V & \longmapsto & (U'_i | i \in I) \end{array}$$

where we used the notation in (7.4.40), (7.4.45) and (7.4.42).

As mentioned earlier, we observe that in view of (7.4.35) - (7.4.40), the domain of definition $B_{[\Omega]}$ of δ as given in (7.4.46), may be *smaller* than the initial tentative one in (7.4.1). On the other hand, the range $A^{[\Omega]}$ of δ in (7.4.46) is certainly *larger* than the tentative one in (7.4.1), since if we define

(7.4.47) $\quad A_{[\Omega]} = \bigcup A_\Delta$

where the union is taken over all nonvoid open $\Delta \subseteq \Omega$, then we have the obvious embedding

(7.4.48) $\quad A_{[\Omega]} \ni U \longmapsto (U) \in A^{[\Omega]}$

where (U) denotes the family which has U alone as an element. Note here the similarity between the above definition of $A_{[\Omega]}$ and that of $B_{[\Omega]}$ in (7.4.40).

This embedding (7.4.48), however, does *not* take us outside of the nowhere dense differential algebras on nonvoid, open subsets of Ω, see (7.4.45).
The effective way for the *computation* of the values of the mapping δ in (7.4.46) will be presented in subsection 7.4.7.

Remark 4

1) The rather unexpected and unusually large range $A^{[\Omega]}$ of δ in (7.4.46), when compared to that attempted intially in (7.4.1), requires a comment about the ways we intend to *use* it. As noted, this range is so large precisely due to the fact that it must *collect* together all the *local* results of the Lie group actions, results which in the general nonprojectable case cannot be obtained globally, as long as we use the nonparametric representation. However, this large range still does not take us out of the nowhere dense differential algebras (1.6), (1.7), (7.1.4). It follows that the only issue we should deal with is that, given $V \in B_{[\Omega]}$, the resulting

$$(7.4.49) \qquad \delta(V) \; = \; (U'_i | i \in I)$$

may happen to be a rather *large family* of generalised functions $U'_i \in A_{\Delta'_i}$, where $\Delta'_i \subseteq \Omega$, with $i \in I$, are nonvoid open subsets, see (7.4.42).

However, let us remember the basic difficulty which the method of parametric Lie group actions, introduced in this work, aims to deal with. Namely, in the classical approach, Olver [1,2], one can deal with arbitrary nonlinear, in particular, nonprojectable Lie group actions on all $\mathcal{C}^\infty$-smooth functions, provided that one does so :

a) *locally,*

b) *one at a time* with respect to both the function and the local domain involved, and

c) not necessarily on the largest possible such local domain.

What is, therefore, missing in such a classical approach is a general enough underlying mathematical structure, which can bring together simultaneously and in a *systematic* manner *all* of the respective local Lie group actions, including, and possibly emphasizing, those which have the largest possible, or maximal, local domains.

The parametric Lie group actions presented in this work can do so, by clearly *isolating* the source of the problem, which as has been mentioned, is *solely* in the difficulties related to the inversion even of one single $\mathcal{C}^\infty$-smooth local diffeomorphism, see the Appendix. The result of this pinpointing is that, in the case of the first track approach, of *nonparametric* representation, the solution of the problem *reduces* to the issue of a proper definition of the mapping δ in (7.4.46). And as mentioned, this definition is *not* at all related to any kind of Lie group actions.

In this way, for a given $V \in B_{[\Omega]}$, we obtain a *gathering together* in $\delta(V) \in A^{[\Omega]}$ of the *deparametrised* results of all the possible local Lie group actions, which have been encapsulated in a *parametric* form in the given V. It follows that, one way to use this abundance of resulting deparametrised local actions in (7.4.49) is to focus - whenever possible, and also convenient - only on some of those resulting actions $U'_i \in A_{\Delta'_i}$, for which their respective domains $\Delta'_i \subseteq \Omega$ may be *larger.* This, of course, amounts to nothing else than to focus on some *subfamilies* in (7.4.49). In other words, we may upon convenience *replace* δ in (7.4.46) with a simpler version of it which, see (7.4.42), only retains certain of its generalised functions $U_i' \in A_{\Delta_i'}$. For instance, as mentioned, we may only be interested in those U_i' which have large, or possibly maximal, domains of definition Δ_i'. This will actually often be done in the sequel.

2) We note that, alternatively, it will be of interest to see when the definition of δ in (7.4.46) may ***simplify*** to a mapping

$$(7.4.50) \qquad B_{[\Omega]} \xrightarrow{\delta} A_{[\Omega]}$$

which, as far as the *range* of δ is concerned, is *identical* with the initially attempted one in (7.4.1), and as such, it *fully recovers* the classical situation, Olver [1,2].

Sometime, like for instance, in the case of projectable Lie group actions, or even of some more general ones, we shall be interested in obtaining a yet more simple form of δ, namely, see (7.1.4)

$$(7.4.51) \qquad B_\Omega \overset{\delta}{\longmapsto} A_\Omega$$

a form which, in view of the usual, nonparametric approach, Olver [1,2], is in fact the ***simplest*** possible one.

Here we present certain conditions, when the simplifications in δ mentioned in (7.4.50), and respectively (7.4.51), can indeed be obtained. The respective proofs, presented in the sequel, will easily follow from (A.31) - (A.39) in the Appendix.

3) A set of conditions for the first of the mentioned kind of simplifications, namely, that in (7.4.50), is as follows. Let a nonvoid and open $\Delta \subseteq \Omega$ be given, and let us take any

$$(7.4.52) \qquad V \in B_\Delta$$

for which there exists, see (7.4.34), (7.4.39)

$$(7.4.53) \qquad V \overset{1}{\longmapsto} (V_\nu | \nu \in \mathbf{N}) \in (\mathcal{C}^\infty(\Delta, M))^{\mathbf{N}}$$

such that, see (7.4.19), (A.32) and pct.2) in Remark 3

$$(7.4.54) \qquad D_\nu \xrightarrow{V_{\nu 1}} V_{\nu 1}(D_\nu), \quad \nu \in \mathbf{N}$$

are $\mathcal{C}^\infty$-smooth *diffeomorphisms*, and in addition, see (7.4.33)

$$(7.4.55) \qquad V_{\nu 1}(D_\nu) = \bar{\Delta}', \quad \nu \in \mathbf{N}$$

for a certain nonvoid open $\bar{\Delta}' \subseteq \Omega$. Then in view of (A.36), the commutative diagrams (7.4.32) can obviously be constructed. Moreover, some of them will be *restrictions* to nonvoid open subsets in $\bar{\Delta}'$ of the following *largest* one among them

(7.4.56)

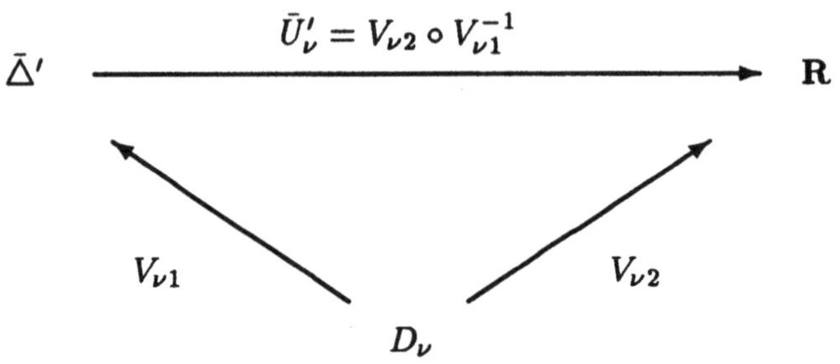

Therefore, in this case, the information given in (7.4.49) contains a certain amount of *redundancy*. Indeed, let $\bar{i} \in I$ be such that

$$(7.4.57) \quad U'_{\bar{i}} = \bar{U}' = \text{class}_{nd}(\bar{U}'_\nu | \nu \in \mathbf{N}) \in A_{\bar{\Delta}'}$$

then a number of temrs corresponding to $i \in I$, $i \neq \bar{i}$ will be but restrictions of $\bar{U}$ to their respective domains Δ'_i.

The effect of (7.4.52) - (7.4.57) is a certain simplification in (7.4.49), for instance, upon *convenience*, we can *limit* ourselves to the following expression of δ

$$(7.4.58) \qquad \delta(V) = \bar{U}' \in A_{\bar{\Delta}'}$$

and we have obtained the *simplification* in (7.4.50). Of course, if in (7.4.53) we would choose another sequence, say $(W_\nu | \nu \in \mathbf{N}) \in (\mathcal{C}^\infty(\Delta, M))^{\mathbf{N}}$, then we may, or may not obtain a corresponding version of (7.4.54) - (7.4.57), with the same large domain as above. Hence, in order to include the generalised functions on other domains, possibly large as well, which may correspond to the sequence $(W_\nu | \nu \in \mathbf{N})$, we would have to go back to (7.4.49), instead of the simplified version of δ in (7.4.58), (7.4.50).

4) A second set of still more particular conditions for simplifying δ to the *best* possible form, given in (7.4.51) - a form, which as we noted, is obtainable in the usual nonparametric approach, Olver [1,2], and in the case of nonprojectable Lie group actions - is presented now. Let us take any

$$(7.4.59) \qquad V \in B_\Omega$$

such that there exists, see (7.4.34), (7.4.39)

$$(7.4.60) \qquad V \overset{1}{\longmapsto} (V_\nu | \nu \in \mathbf{N}) \in (\mathcal{C}^\infty(\Omega, M))^{\mathbf{N}}$$

for which, see (A.37), (A.39)

$$(7.4.61) \qquad \Omega \overset{V_{\nu 1}}{\longmapsto} V_{\nu 1}(\Omega), \quad \nu \in \mathbf{N}$$

are $\mathcal{C}^\infty$-smooth *diffeomorphisms*, and in addition

$$(7.4.62) \qquad V_{\nu 1}(\Omega) = \Omega, \quad \nu \in \mathbf{N}$$

In this case, under a comment similar to that related to (7.4.58) above, that relation becomes

(7.4.63) $$\delta(V) = \bar{U}' \in A_\Omega$$

and we have obtained the simplification in (7.4.51).

5) As expected, in the case of *projectable* Lie group actions, see (7.4.72) - (7.4.77) below, the conditions (7.4.59) - (7.4.62) are trivially satisfied.
It should be mentioned that in Examples 4 and 5 in subsection 7.5.3, the conditions (7.4.59) - (7.4.62) are as well satisfied in the case of certain *nonprojectable* Lie group actions.

In (7.5.14) - (7.5.16) below, we can see an example when, again, conditions (7.4.52) - (7.4.57) are satisfied in the case of possibly nonprojectable Lie group actions.

For further comment on the definition of δ given in (7.4.46), see below Remark 5 in subsection 7.5.1.

6) The issue of the possible simplification of the mapping δ proves to be more complicated than suggested in pct. 2) - 4) above, see details in this respect in Example 5 and Remark 8 in section 7.5.
This is in particular true with respect to what it means a *maximal* instance of a generalised function in the family, see (7.4.46), (7.4.42), which, in the general situation, gives the value of δ.

4.5. Constructing Arbitrary Lie Group Actions on Generalised Functions

We recapitulate. The idea which we want to use for defining the arbitrary nonlinear Lie group actions, see (7.3.12), (7.1.11)

(7.4.64) $$G \times ((\mathcal{C}^\infty_{par}(\Omega))^{\mathbf{N}} / \approx_{nd}) \xrightarrow{\mathcal{L}} (\mathcal{C}^\infty_{par}(\Omega))^{\mathbf{N}} / \approx_{nd}$$

is to do it through a commutative diagram

(7.4.65) $$\begin{array}{ccc} G \times ((\mathcal{C}^\infty_{par}(\Omega, M))^{\mathbf{N}} / \approx_{nd}) & \xrightarrow{\mathcal{L}_{prm}} & (\mathcal{C}^\infty_{par}(\Omega, M))^{\mathbf{N}} / \approx_{nd} \\ \uparrow \mathrm{id}_G \times \beta_* & & \downarrow \delta \\ G \times ((\mathcal{C}^\infty_{par}(\Omega))^{\mathbf{N}} / \approx_{nd}) & \xrightarrow[\mathcal{L}]{} & (\mathcal{C}^\infty_{par}(\Omega))^{\mathbf{N}} / \approx_{nd} \end{array}$$

in which, in view of (7.3.13), (7.3.14), only the mapping δ has presented us with the problem of its appropriate definition, a problem solved in this subsection, in (7.4.46). And as mentioned earlier, this problem does not involve any kind of Lie group action. Its only difficulties, see Appendix, come from the rather inevitable and natural, although somewhat involved, ways one is able, or rahter obliged, to invert into a $\mathcal{C}^\infty$-smooth function a given $\mathcal{C}^\infty$-smooth local diffeomorphism which, therefore, may fail to be injective.

Now, owing to (7.3.15), (7.4.1), the precise definition of δ obtained at last in (7.4.46) implies a change in (7.4.64), and therefore, in (7.4.65) as well. This is due to the fact that, as seen in (7.4.40), the domain of definition of δ given in (7.4.46) may turn out to be *smaller* than intially attempted in (7.3.15), (7.4.1), and thus also in (7.4.65). At the same time, we may expect changes in (7.4.64) and (7.4.65) owing to the fact that the range of δ in (7.4.46) is certainly *larger* than that attempted initially in the mentioned relations.

It follows that, so far, the initially attemtped commutative diagram (7.4.65) will have to take the form

(7.4.66)
$$\begin{array}{ccc} G \times C_{[\Omega]} & \xrightarrow{\ \mathcal{L}_{prm}\ } & B_{[\Omega]} \\ \Big\uparrow \mathrm{id} \times \beta_* & & \Big\downarrow \delta \\ G \times D_{[\Omega]} & \xrightarrow[\ \mathcal{L}\ ]{} & A^{[\Omega]} \end{array}$$

where all the mappings are by now well defined, and where one is only left with the task to suitably determine the spaces, see (7.3.17), (7.4.65), (7.4.47)

(7.4.67)
$$C_{[\Omega]} \subseteq (\mathcal{C}^\infty_{par}(\Omega, M))^{\mathbf{N}} / \approx_{nd}$$
$$D_{[\Omega]} \subseteq (\mathcal{C}^\infty_{par}(\Omega))^{\mathbf{N}} / \approx_{nd} \; = \; A_{[\Omega]}$$

These two spaces will be defined in subsection 7.6, where a commutative diagram containing (7.4.66) as a particular case will be constructed, see (7.6.3).

4.6. The Particular Case of Projectable Lie Group Actions

Before going further, let us show that the arbitrary nonlinear Lie group actions (7.4.66) contain as a *particular* case the *projectable* Lie group actions on the nowhere dense differential algebras (1.6), (7.1.4), actions which were constructed in Oberguggenberger & Rosinger, Rosinger [11], Rosinger & Rudolph, and Rosinger & Walus [1,2], see also chapter 10. Moreover, we shall show that in the case of the action of projectable Lie groups, the mapping δ in (7.4.46) can - upon convenience - be *simplified* to the form in (7.4.51), (7.4.63), since the conditions (7.4.59) - (7.4.62) will automatically be satisfied. Indeed, let us take any

$$(7.4.68) \qquad U = \text{class}_{nd}(U_\nu | \nu \in \mathbf{N}) \in (\mathcal{C}^\infty(\Omega))^{\mathbf{N}} / \approx_{nd}$$

then, see (7.3.16)

$$\beta_*(U) = \text{class}_{nd}(U_{\nu *} | \nu \in \mathbf{N}) \in (\mathcal{C}^\infty(\Omega, M))^{\mathbf{N}} / \approx_{nd}$$

Therefore, given $g \in G$, we have, see (7.3.11)

$$(7.4.69) \qquad g\, \beta_*(U) = \text{class}_{nd}(gU_{\nu *} | \nu \in \mathbf{N}) \in (\mathcal{C}^\infty(\Omega, M))^{\mathbf{N}} / \approx_{nd}$$

Let us now take

$$(7.4.70) \qquad V = g\, \beta_*(U)$$

and compute $\delta(V)$. For that, we shall take

$$V \stackrel{1}{\longmapsto} (V_\nu | \nu \in \mathbf{N})$$

in (7.4.60), and do so in a special way, as suggested by (7.4.70) and (7.4.69), namely

$$(7.4.71) \qquad V_\nu = g\, U_{\nu *}, \quad \nu \in \mathbf{N}$$

However, we assumed that we are dealing with a *projectable* Lie group action, therefore (2.3), (7.4.71), (7.4.2) yield

$$(7.4.72) \qquad V_{\nu 1}(x) = g_1(x), \quad \nu \in \mathbf{N},\ x \in \Omega$$

hence according to (2.4), the mappings

$$(7.4.73) \qquad \Omega \xrightarrow{V_{\nu 1}} g_1(\Omega) = \Omega, \quad \nu \in \mathbf{N}$$

being all identical with g_1, will all be $\mathcal{C}^\infty$-smooth diffeomorphisms. Furthermore, (7.4.18) implies

$$D_\nu = \Omega, \quad \nu \in \mathbf{N}$$

while (7.4.22) gives

$$\mathcal{D}_\nu = \{\ \Delta''_\nu \subseteq \Omega \mid \Delta''_\nu \text{ nonvoid, open }\}, \quad \nu \in \mathbf{N}$$

In this way, if we recall (7.4.36) - (7.4.38) as well, it follows that

$$(7.4.74) \qquad V \in B_\Omega$$

and therefore (7.4.59) - (7.4.62) are indeed satisfied, thus (7.4.63) does hold and, under comments similar to those related to it, it will give us

(7.4.75) $\delta(V) = U' \in A_\Omega$

where

(7.4.76) $U' = \text{class}_{nd}(U'_\nu | \nu \in \mathbf{N})$

with

(7.4.77) $U'_\nu(x') = g_2(g_1^{-1}(x'), U'_\nu(g_1^{-1}(x'))), \quad \nu \in \mathbf{N}, \ x' \in \Omega$

4.7. The Computation of δ

In order to better understand some of the *significant* differences between the usual, nonparametric, and on the other hand, parametric approaches, it will be useful to present here in some detail the effective way one can compute the *value* of the mapping δ in (7.4.46), that is, the way to determine the whole *family* of generalised functions U'_i, with $i \in I$.
Indeed, it will be seen that, when pursuing the first track, that is, the usual, nonparametric approach, such a computation of δ is the unavoidable *last step*. And although this computation in itself does not involve any Lie group action, it nevertheless leads to significant technical complications, owing merely to the fact that, at the end, one wants to obtain a usual, nonparametric generalised function, a desire which typically, will actually lead to a large family of such functions, each of them defined only locally.
On the other hand, when pursuing the second track approach, one will simply *no longer* need to compute δ, thus avoiding all of the mentioned technical complications.

We recall that the main aim of the present section 7.4 has been to obtain the precise definition of the mapping δ. And as we have already noted, this mapping δ brings in certain complications, both with respect to its domain of definition $B_{[\Omega]}$, and its range $A^{[\Omega]}$, none of these complications, however, being related to the actual Lie group actions.
In Remark 3, we have addressed some issues related to the domain of δ, while in Remark 4, we have done the same with respect to its range, and in particular, its values.
Here, we shall summarize the way the *computation* of the value of δ, that is, the determination of the family of generalised functions, see (7.4.46)

(7.4.78) $\delta(V) = (U'_i \mid i \in I)$

for a given $V \in B_{[\Omega]}$, can be done.

Let us note that, in view of the structure of $B_{[\Omega]}$ in (7.4.40), it suffices to assume that

(7.4.79) $V \in B_\Delta$

for a certain nonvoid open subset $\Delta \subseteq \Omega$, in which case the relations (7.4.36) - (7.4.38) apply. Therefore, to compute $\delta(V)$ means to compute, see (7.4.42)

(7.4.80) $\quad V \overset{5}{\longmapsto} (\, (\Delta'_i, U'_i) \mid i \in I\,)$

Here, the nonvoid open subsets $\Delta'_i \subseteq \Omega$ and the generalised functions $U'_i \in A_{\Delta'_i}$, with $i \in I$, are supposed to *exhaust* all such pairs (Δ'_i, U'_i) which satisfy (7.4.38). This means that for each such pair, with $i \in I$ given, we have the property

$\forall\, x' \in \Delta'_i$:

$\exists\, (V_\nu | \nu \in \mathbf{N}) \in (\mathcal{C}^\infty(\Delta, M))^{\mathbf{N}},\ x' \in \Delta^0 \subseteq \Delta'_i,\ \Delta^0$ open, $U^0 \in A_{\Delta^0}$:

1) $V \;=\; \text{class}_{nd}(V_\nu | \nu \in \mathbf{N})$

2) the commutative diagrams of $\mathcal{C}^\infty$-smooth functions hold for $\nu \in \mathbf{N}$

(7.4.81)

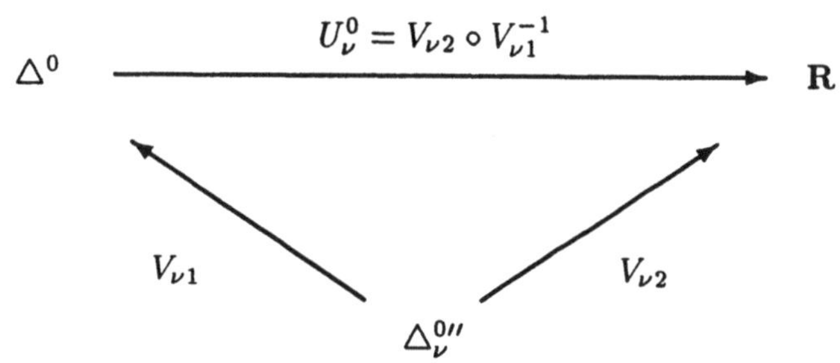

with

$\Delta^0 = V_{\nu 1}(\Delta^{0\prime\prime}_\nu),\ \Delta^{0\prime\prime}_\nu \in \mathcal{D}_{V_\nu}$

3) $U^0 \;=\; \text{class}_{nd}(U^0_\nu | \nu \in \mathbf{N})$

4) $U'_i \;=\; U^0$ on Δ^0

It follows that the computation of $\delta(V)$ simply means to *collect* all the possible such pairs (Δ'_i, U'_i) which satisfy (7.4.81).
Here we show how for a given, see (7.4.79)

(7.4.82) $\quad V \in B_\Delta$

we can find *all* the pairs (Δ^0, U^0) which appear in (7.4.81). Let us take any $(V_\nu | \nu \in \mathbf{N}) \in (\mathcal{C}^\infty(\Delta, M))^{\mathbf{N}}$, such that, see pct. 1) in (7.4.81)

(7.4.83) $\quad V \;=\; \text{class}_{nd}(V_\nu | \nu \in \mathbf{N})$

and let us define, see (7.4.22)

(7.4.84) $\quad \mathcal{D}^0_{(V_\nu | \nu \in \mathbf{N})} \;=\; \liminf_{\nu \in \mathbf{N}}\ V_{\nu 1}(\mathcal{D}_{V_\nu})$

which is a possibly void set of nonvoid open subsets in Ω.
Now assuming that

(7.4.85) $\mathcal{D}^0_{(V_\nu|\nu\in\mathbf{N})} \neq \phi$

let us take any

(7.4.86) $\Delta^0 \in \mathcal{D}^0_{(V_\nu|\nu\in\mathbf{N})}$

which is therefore a nonvoid open subset of Ω. Then (7.4.84) implies that, see pct. 2) in (7.4.81)

$$(7.4.87)\quad \begin{array}{l} \exists\ \mu \in \mathbf{N}\ : \\ \forall\ \nu \in \mathbf{N},\ \nu \geq \mu\ : \\ \exists\ \Delta^{0\prime\prime}_\nu \in \mathcal{D}_{V_\nu}\ : \\ \qquad \Delta^{0\prime\prime}_\nu \xrightarrow{V_{\nu 1}} \Delta^0 \quad \text{is a } \mathcal{C}^\infty\text{-smooth diffeomorphism} \end{array}$$

In this way, with the use of the commutative diagram at pct. 2) in (7.4.81), we can construct the generalised function

(7.4.88) $U^0 = \text{class}_{nd}(U^0_\nu|\nu \in \mathbf{N}) \in A_{\Delta^0}$

It follows that (7.4.83), (7.4.86) - (7.4.88) give us all the pairs (Δ^0, U^0) which appear in (7.4.81) for a given $V \in B_\Delta$ and Δ'_i in (7.4.79), (7.4.80). And then, we obtain the *lower bound* on the value of $\delta(V)$, given by

$$(7.4.89)\quad \delta(V) \supseteq \left(U^0 \;\middle|\; \begin{array}{l} \Delta^0 \in \mathcal{D}^0_{(V_\nu|\nu\in\mathbf{N})} \\ \\ V = \text{class}_{nd}(V_\nu|\nu \in \mathbf{N}) \end{array} \right)$$

where the inclusion $\supseteq$ means that the right hand term is a *subfamily* of the left hand term, when the latter is considered in its full generality in (7.4.48).

We note that this lower bound is twice relevant : first, it is completely *general*, that is, it holds without any restrictions, except of course the one in (7.4.82), and second, it shows how *large* the family which gives the value of $\delta(V)$ can be, unless we try to limit ourselves to certain maximal elements of it. Here, maximality proves itself to be a rather complex notion, not being reduced to the mere mximality of the domain of a generalised function, see for instance Example 5 in subsection 7.5.3 for further details.

From (7.4.89) however, it is rather more difficult to go over, and determine all the pairs (Δ'_i, U'_i), with $i \in I$, in (7.4.80), which through (7.4.78) give the actual value of $\delta(V)$. Indeed, as far as the nonvoid open subsets $\Delta'_i \subseteq \Omega$ are concerned, in view of (7.4.81), it is clear that each of them is a union of certain nonvoid open subsets Δ^0 in (7.4.86). However, the determination of the corresponding generalised functions $U'_i \in A_{\Delta'_i}$ from the generalised functions $U^0 \in A_{\Delta^0}$ is a less straightforward process.

Let us show the usefulness of the lower bound in (7.4.89) on the value of $\delta(V)$. In view of (7.4.84), it is clear that conditions (7.4.54), (7.4.55) imply

$$\bar{\Delta}' \in \mathcal{D}^0_{(V_\nu | \nu \in \mathbf{N})}$$

therefore, (7.4.58) is a direct consequence of (7.4.89).
The same, of course, applies to the more particular situations in (7.4.61) - (7.4.63), or in (7.4.72) - (7.4.75).

We give another application of the lower bound in (7.4.89). Let us assume that in (7.4.82), (7.4.83) we have the situation, when

$$(7.4.90) \quad V_{\nu 1}(\Delta) \; = \; \Delta', \quad \nu \in \mathbf{N}$$

where $\Delta' \subseteq \Omega$ is nonvoid, open. Now, for given $\nu \in \mathbf{N}$, the closed subset of Δ, see (7.4.18)

$$(7.4.91) \quad \Sigma_{V_\nu} \; = \; \Delta \setminus D_{V_\nu}$$

can be arbitrarily large. For instance, in extreme, it can be the whole of Δ itself. However, according to Sard's theorem, Kahn, we have for $\nu \in \mathbf{N}$

$$(7.4.92) \quad \text{mes}(V_{\nu 1}(\Sigma_{V_\nu})) = 0$$

Let us denote

$$(7.4.93) \quad \Sigma'_{(V_\nu | \nu \in \mathbf{N})} \; = \; \cup_{\nu \in \mathbf{N}} V_{\nu 1}(\Sigma_{V_\nu})$$

then (7.4.92) yields

$$(7.4.94) \quad \text{mes}\; (\Sigma'_{(V_\nu | \nu \in \mathbf{N})}) = 0$$

however, $\Sigma'_{(V_\nu | \nu \in \mathbf{N})}$ need *not* be closed in Δ'. Therefore, it need not be nowhere dense in Δ'.

On the other hand, in view of (7.4.94), we obtain

$$(7.4.95) \quad \Sigma'_{(V_\nu | \nu \in \mathbf{N})} \;\; \text{closed} \iff \Sigma'_{(V_\nu | \nu \in \mathbf{N})} \;\; \text{closed and nowhere dense}$$

Therefore, let us assume that

$$(7.4.96) \quad \Sigma'_{(V_\nu | \nu \in \mathbf{N})} \;\; \text{is closed}$$

then

$$(7.4.97) \quad \Delta'_{(V_\nu | \nu \in \mathbf{N})} \; = \; \Delta' \setminus \Sigma'_{(V_\nu | \nu \in \mathbf{N})} \;\; \text{is open}$$

and clearly, we have

$$\forall\ x' \in \Delta'_{(V_\nu|\nu\in\mathbf{N})},\ \mu \in \mathbf{N}\ :$$

$$(7.4.98) \qquad \exists\ x' \in \Delta^0_\mu \subseteq \Delta'_{(V_\nu|\nu\in\mathbf{N})},\ \Delta^0_\mu \ \text{ open }\ :$$

$$\Delta^0_\mu \xrightarrow{V^{-1}_{\mu 1}} \Delta^{0\prime\prime}_\mu = V^{-1}_{\mu 1}(\Delta^0_\mu) \quad \text{is a } \mathcal{C}^\infty\text{-smooth diffeomorphism}$$

We recall now (7.4.86), (7.4.87), and note that in case in (7.4.98) we have the following *uniformity* with respect to $\mu \in \mathbf{N}$

$$(7.4.99) \qquad \Delta^0_\mu = \Delta^0, \quad \mu \in \mathbf{N}$$

where Δ^0 is a nonvoid open subset in Ω, then

$$(7.4.100) \qquad \Delta^0 \in \mathcal{D}^0_{(V_\nu|\nu\in\mathbf{N})}$$

and we are back to the situation in (7.4.86) - (7.4.89), and thus we obtained one of the pairs (Δ^0, U^0) in (7.4.81), which means that the generalised function $U^0 \in A_{\Delta^0}$ belongs to the family of generalised functions which give the value of $\delta(V)$, that is, see (7.4.78), (7.4.89)

$$(7.4.101) \qquad \delta(V) \supseteq (\ U^0\ )$$

4.8. Extension of the Family Giving the Values of δ

In certain applications, see Example 5, in subsection 7.5.3., or chapter 8, it will be useful to employ a variant of δ defined in (7.4.46) which has the family giving its value somewhat *extended*. This is obtained by relaxing the conditions (7.4.32) and (7.4.33) as follows.

Let us recall here that, as mentioned in Remark 4, in subsection 7.4.4., already the initial definition of δ, see (7.4.46), (7.4.49), leads to what may appear to be an unusually large family for its values. In this way, one may at first be less interested in the kind of further extension which will be implemented here.

Nevertheless, as we have seen, this often large family which gives the values of δ is due to the *fragmentation* introduced by nonprojectable Lie group actions, when we deal with them along the first track, nonparametric approach. And also as seen earlier, we may often be interested to find the maximal pieces in this fragmentation. And then, as seen in Example 5, in subsection 7.5.3., as well as in chapter 8, the extension of the family of values of δ which we present here can help in finding such maximal components.

Let us now proceed with this extension. First, in the commutative diagrams (7.4.32) we will allow that the nonvoid open subsets Δ' may depend on $\nu \in \mathbf{N}$, provided that certain conditions specified below are satisfied. In other words, the commutative diagrams (7.4.32) will, for $\nu \in \mathbf{N}$, be replaced with the following less restrictive ones

(7.4.102)
$$\begin{array}{ccc} \Delta'_\nu & \xrightarrow{\quad U'_\nu = V_{\nu 2} \circ V_{\nu 1}^{-1} \quad} & \mathbf{R} \\ & {}_{V_{\nu 1}} \nwarrow \qquad \nearrow {}_{V_{\nu 2}} & \\ & \Delta''_\nu & \end{array}$$

where instead of (7.4.33), we now have

(7.4.103) $\quad \Delta'_\nu = V_{\nu 1}(\Delta''_\nu), \quad \Delta''_\nu \in \mathcal{D}_\nu = \mathcal{D}_{V_\nu}$

and furthermore, there exists a nonvoid open subset $\Delta' \subseteq \Omega$, such that

(7.4.104)
$$\Delta'_0 \subset \bar{\Delta}'_0 \subset \Delta'_1 \subset \ \ldots \ \subset \Delta'_\nu \subset \bar{\Delta}'_\nu \subset \Delta'_{\nu+1} \subset \ \ldots \ \Delta'$$
$$\Gamma' = \Delta' \setminus \bigcup_{\nu \in \mathbf{N}} \Delta'_\nu \quad \text{is closed and nowhere dense in} \quad \Delta'$$

In view of the extension in (7.4.102) - (7.4.104), we have to make a corresponding extension in the multivalued mapping 2 in (7.4.34) as well. And obviously, this can be done as follows

(7.4.105) $\quad (V_\nu | \nu \in \mathbf{N}) \xrightarrow{2} \Delta' \ \text{ in } \ (7.4.104)$

Now we note that in view of the commutative diagrams (7.4.102), we still can construct the sequence of functions

(7.4.106) $\quad U'_\nu = V_{\nu 2} \circ V_{\nu 1}^{-1} \in \mathcal{C}^\infty(\Delta'_\nu), \quad \nu \in \mathbf{N}$

however, since they need no longer have the same domain of definition Δ', we cannot use them to construct the multivalued mapping 3 in (7.4.34).
Nevertheless, we can overcome this difficulty by suitably *extending* the functions in (7.4.105), using the above properties (7.4.103), (7.4.104). In this way, we shall construct a sequence of functions associated with (7.4.102), namely

(7.4.107) $\quad (U^\#_\nu | \nu \in \mathbf{N}) \in (\mathcal{C}^\infty(\Delta'))^{\mathbf{N}}$

such that

(7.4.108)
$$\begin{array}{l} \forall \ x' \in \Delta' \setminus \Gamma' \ : \\ \exists \ \Delta^\# \subseteq \Delta' \setminus \Gamma' \ \text{ neighbourhood of } \ x', \ \ \mu \in \mathbf{N} \ : \\ \forall \ \nu \in \mathbf{N}, \ \nu \geq \mu \ : \\ \quad U^\#_\nu = U'_\nu \quad \text{on} \ \ \Delta^\# \end{array}$$

Clearly, in view of (7.4.108), (7.1.1), we shall now be able to define the multivalued mapping 3 in (7.4.34) by

$$(7.4.109) \quad \Delta' \text{ in } (7.4.104) \xrightarrow{3} U^{\#} = \text{class}_{nd}(U_\nu^{\#} | \nu \in \mathbf{N}) \in A_{\Delta'}$$

With these modifications, more precisely, extensions of (7.4.34), it is clear that we can still go through the stages (7.4.35) - (7.4.46), and thus arrive at the intended extension of the family giving the values of δ, namely

$$(7.4.110) \quad \delta(V) \supseteq (\ U^{\#}\ )$$

Now we shall show the construction of the sequence of functions (7.4.107) which has the property (7.4.108).

This construction follows from a quite general method applicable in the case of the nowhere dense differential algebras of generalised functions A_Ω, and it will also be presented in its somewhat different details in the important particular situation of (7.5.30) - (7.5.48).

Here, we use it in the following way :

Let us be given the nonvoid open subsets $\Delta_\nu \subset \Omega$, with $\nu \in \mathbf{N}$, such that, see (7.4.104)

$$(7.4.111) \quad \begin{array}{l} \Delta_0 \subset \bar{\Delta}_0 \subset \Delta_1 \subset \ \ldots \ \subset \Delta_\nu \subset \bar{\Delta}_\nu \subset \Delta_{\nu+1} \subset \ \ldots \ \subset \Omega \\ \Gamma = \Omega \setminus \bigcup_{\nu \in \mathbf{N}} \Delta_\nu \quad \text{is closed and nowhere dense in} \quad \Omega \end{array}$$

Further, suppose given any sequence of functions

$$(7.4.112) \quad U_\nu \in \mathcal{C}^\infty(\Delta_\nu), \quad \nu \in \mathbf{N}$$

Then there exists a *unique* $U \in A_\Omega$, such that for a suitable representation of it

$$(7.4.113) \quad U = \text{class}_{nd}(U_\nu^{\#} | \nu \in \mathbf{N}), \quad (U_\nu^{\#} | \nu \in \mathbf{N}) \in (\mathcal{C}^\infty(\Omega))^{\mathbf{N}}$$

we shall have

$$(7.4.114) \quad \begin{array}{l} \forall\ x \in \Omega \setminus \Gamma\ : \\ \exists\ \Delta \subseteq \Omega \setminus \Gamma \ \text{ neighbourhood of } x, \ \mu \in \mathbf{N}\ : \\ \forall\ \nu \in \mathbf{N},\ \nu \geq \mu\ : \\ \quad U_\nu^{\#} = U_\nu \quad \text{on} \ \Delta \end{array}$$

Indeed, in view of (7.4.111), for $\nu \in \mathbf{N}$, we have $\Omega \setminus \Delta_{\nu+1} \subset \Omega \setminus \bar{\Delta}_\nu$, with the first subset being nonvoid closed, while the second being open. Therefore, Kahn, we can take functions $\gamma_\nu', \gamma_\nu'' \in \mathcal{C}^\infty(\Omega)$, such that

$$\gamma'_\nu \geq 0 \quad \text{on } \Omega$$

$$\Omega \setminus \Delta_{\nu+1} = \{x \in \Omega \mid \gamma'_\nu(x) = 0\}$$

(7.4.115) $$0 \leq \gamma''_\nu \leq 1 \quad \text{on } \Omega$$

$$\gamma''_\nu = \left| \begin{array}{l} 1 \quad \text{on } \Omega \setminus \Delta_{\nu+1} \\ \\ 0 \quad \text{on } \bar{\Delta}_\nu \end{array} \right.$$

Then the function $\gamma_\nu \in \mathcal{C}^\infty(\Omega)$ given by

(7.4.116) $$\gamma_\nu = \gamma'_\nu + 1 - \gamma''_\nu$$

is such that

$$\gamma_\nu \geq 0 \quad \text{on } \Omega$$

(7.4.117) $$\gamma_\nu \geq 1 \quad \text{on } \bar{\Delta}_\nu$$

$$\Omega \setminus \Delta_{\nu+1} = \{x \in \Omega \mid \gamma_\nu(x) = 0\}$$

Finally, we take any function $b \in \mathcal{C}^\infty(\mathbf{R})$, such that

(7.4.118) $$b = \left| \begin{array}{l} 0 \quad \text{on } [-1/2, 1/2] \\ \\ 1 \quad \text{on } \mathbf{R} \setminus (-1, 1) \end{array} \right.$$

And now, for $\nu \in \mathbf{N}$, we can define $U^\#_\nu \in \mathcal{C}^\infty(\Omega)$ by

(7.4.119) $$U^\#_\nu(x) = \left| \begin{array}{ll} b((\nu+1)\gamma_\nu(x))U_\nu(x) & \text{if } x \in \Delta_{\nu+1} \\ \\ 0 \quad \text{if } x \in \Omega \setminus \Delta_{\nu+1} & \end{array} \right.$$

and the property (7.4.114) will easily follow from (7.4.115) - (7.4.119). The uniqueness of $U \in A_\Omega$ is implied by (7.1.1) and (7.4.113), (7.4.114).

5. Two Basic Applications

Before determining the two spaces $C_{[\Omega]}$ and $D_{[\Omega]}$, in (7.4.66), (7.4.67), and in order to *both test and better understand* what has actually been involved in the construction of δ, and hence, of (7.1.11), (7.3.12), (7.4.1), (7.4.66) and (7.4.67), let us first apply the above construction in *two* of the most relevant cases of its use in applications of Lie group actions.

Namely, we shall start by aplying the above construction when arbitrary nonlinear, including nonprojectable, Lie groups act on a *usual*, that is, non parametrically given $\mathcal{C}^\infty$-smooth function.

Second, we shall turn to such arbitrary Lie groups acting on a large class of *generalised functions* in the nowhere dense differential algebras. This large class is *not* contained in the distributions of Schwartz, see (7.1.10), *nor* in the generalised functions in the sense of Colombeau.
The interest in this class of generalised functions comes from the fact that it contains the *global* solutions to *arbitrary nonlinear analytic* partial differential equations, see Rosinger [7-9], Rosinger & Walus [1,2]. In subsection 7.5.2, the necessary basics about the respective result on the existence of generalised solutions will shortly be reviewed.

5.1. Arbitrary Lie Group Actions on $\mathcal{C}^\infty$-smooth Functions

Let us start with the *first test*, and consider the *classical* case, when the commutative diagram (7.4.66) is supposed to act on an arbitrary $\mathcal{C}^\infty$-smooth function

$$(7.5.1) \qquad \psi : \Delta \longrightarrow \mathbf{R}$$

where $\Delta \subseteq \Omega$ in nonvoid, open. Then in view of (7.1.6), (7.4.47), we have the embedding

$$(7.5.2) \qquad \psi \longmapsto U = u(\psi) + \mathcal{I}_{nd}(\Delta) \in A_\Delta \subseteq A_{[\Omega]}$$

Applying now the mapping (7.3.16) to U, we obtain, see also (7.3.17)

$$(7.5.3) \qquad \beta_*(U) \ = \ \text{class}_{nd}\ u(\psi_*) \in (\mathcal{C}^\infty(\Delta, M))^{\mathbf{N}} / \approx_{nd}$$

where, see (3.5)

$$\Delta \ni x \longmapsto \psi_*(x) = (x, \psi(x)) \in M$$

Let us now take any group element $g \in G$, then in view of (7.5.3), (7.3.11), it follows that

$$(7.5.4) \qquad g\ \beta_*(U) \ = \ \text{class}_{nd} u(g\ \psi_*) \in (\mathcal{C}^\infty(\Delta, M))^{\mathbf{N}} / \approx_{nd}$$

where according to (4.2), (1.2), we have

$$(7.5.5) \quad \Delta \ni x \ \longmapsto \ (g\ \psi_*)(x) = (g_1(x, \psi(x)), g_2(x, \psi(x))) \in M$$

The result which is of importance to us is that, see (7.4.40), (7.4.66), and also (5.8)

$$(7.5.6) \qquad g\ \beta_*(U) \in B_\Delta \subseteq B_{[\Omega]}, \quad \text{if } \Delta_{(g,\psi)} \neq \phi$$

In order to prove (7.5.6), we recall pct.2) in Remark 3. In particular, we turn to (7.4.36) - (7.4.38), and in view of (7.5.4), we choose

$$(7.5.7) \quad V \ = \ g\ \beta_*(U), \quad (V_\nu | \nu \in \mathbf{N}) \ = \ u(g\ \psi_*)$$

Then (7.5.5) gives for $x \in \Delta$, $\nu \in \mathbf{N}$

$$V_{\nu 1}(x) = g_1(x, \psi(x))$$

$$V_{\nu 2}(x) = g_2(x, \psi(x))$$

Let us define the following two mappings α, see (2.11), and α'

$$(7.5.8) \qquad \begin{aligned} &\Delta \ni x \stackrel{\alpha}{\longmapsto} g_1(x, \psi(x)) \in \Omega \\ &\Delta \ni x \stackrel{\alpha'}{\longmapsto} g_2(x, \psi(x)) \in \mathbf{R} \end{aligned}$$

and let us assume that, see (5.7), (5.8), (5.26) - (5.30)

$$(7.5.9) \qquad \Delta_{(g,\psi)} = D_\alpha = \{x \in \Delta \mid J_\alpha(x) \neq 0\} \neq \phi$$

Then the mappings (7.4.19) become for $\nu \in \mathbf{N}$, one and the same $\mathcal{C}^\infty$-smooth surjective local diffeomorphism, namely

$$D_\nu = D_\alpha \stackrel{V_{\nu 1} = \alpha}{\longrightarrow} V_{\nu 1}(D_\nu) = \alpha(D_\alpha)$$

and the commutative diagrams (7.4.32), (7.4.81) will take, for $\nu \in \mathbf{N}$, the following identical form

$$(7.5.10) \qquad \begin{array}{ccc} \Delta' & \xrightarrow{\quad U'_\nu = \alpha' \circ \alpha^{-1} \quad} & \mathbf{R} \\ & \nwarrow \; V_{\nu 1} = \alpha \qquad V_{\nu 2} = \alpha' \; \nearrow & \\ & \Delta'' & \end{array}$$

where, see (7.4.22), (A.5), (7.5.8)

$$\Delta'' \in \mathcal{D}_\nu = \mathcal{D}_\alpha \neq \phi, \qquad \Delta' = \alpha(\Delta'') \text{ open}$$

Based on the above, for each Δ', we define the generalised function, see also pct. 3) in (7.4.81)

$$(7.5.11) \qquad U' = \text{class}_{nd}(U'_\nu | \nu \in \mathbf{N}) = \text{class}_{nd} u(\alpha' \circ \alpha^{-1}) \in A_{\Delta'}$$

It follows that in the case of (7.5.1) - (7.5.4), and thus of (7.5.7), we are back to the situation in (7.4.13), (7.4.14). In particular, (7.4.39) holds with the choice of $(V_\nu | \nu \in \mathbf{N})$ in (7.5.7), and for Δ', U' in (7.5.10). In this way, (7.4.42) becomes

(7.5.12) $\quad V \overset{5}{\longmapsto} ((\Delta', U') \mid \Delta'' \in \mathcal{D}_\alpha)$

that is $I = \mathcal{D}_\alpha$, $i = \Delta''$, while Δ' and $U' \in A_{\Delta'}$ will correspond to Δ'' according to (7.5.9) - (7.5.11).

Now in view of pct.2) in Remark 3, we shall obtain (7.5.6), while at the same time (7.4.46), (7.5.12) give

(7.5.13) $\quad \delta(g\ \beta_*(U)) \supseteq (U' \mid \Delta'' \in \mathcal{D}_\alpha)$

The Classical Case. It is useful to consider the yet more particular situation when in the classical case of (7.5.1) - (7.5.4), the result of the group action $g\ \psi$ of $g \in G$ on the $\mathcal{C}^\infty$-smooth function $\psi : \Delta \to \mathbf{R}$ is *again* a well defined $\mathcal{C}^\infty$-smooth function $\varphi : \bar{\Delta} \to \mathbf{R}$, in other words, $g\ \psi$ can be defined *globally*, and it will be given by the function φ.
This happens, see (2.11) - (2.13), when for instance, the mapping

(7.5.14) $\quad \Delta \ni x \overset{\alpha}{\longmapsto} g_1(x, \psi(x)) \in \bar{\Delta} = \alpha(\Delta)$

is in fact a $\mathcal{C}^\infty$-smooth *diffeomorphism*, although the respective Lie group action need not necessarily be projectable, see Examples 4 and 5 in subsection 7.5.3 below. Then, as seen in (2.12), we shall obtain φ given by

(7.5.15) $\quad \bar{\Delta} \ni \bar{x} \overset{\varphi}{\longmapsto} g_2(\alpha^{-1}(\bar{x}), \psi(\alpha^{-1}(\bar{x}))) \in \mathbf{R}$

In this case, the identical commutative diagrams (7.5.10) will become *restrictions* to nonvoid open subsets in $\bar{\Delta}$ of the following *largest* one among them

(7.5.16)

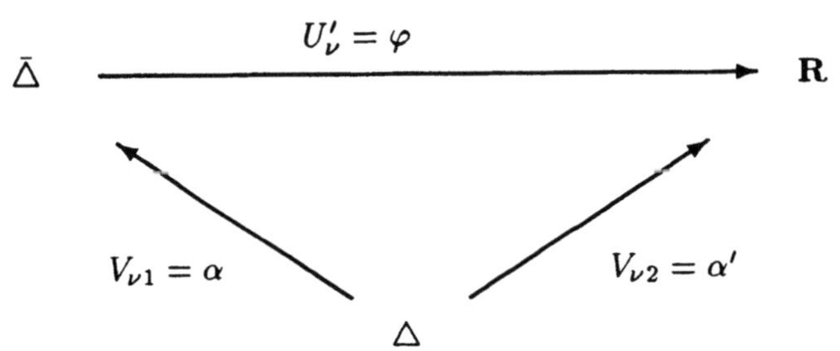

Therefore, as in (7.4.50), (7.4.58), we can now *simplify* δ in (7.5.13), by writing its value as the *one term* family

(7.5.17) $\quad \delta(g\ \beta_*(U)) \ = \ \text{class}_{nd}u(\varphi) \in A_{\bar{\Delta}}$

and in this way we can again recover the classical result, Olver [1,2], through the use of our general construction in (7.4.46), see Remark 4 for comments on such a simplification.

As noted earlier, (7.5.14) will hold for projectable Lie group actions, hence in such a case (7.5.15) - (7.5.17) will also be valid.

However, as seen in Examples 4 and 5 in subsection 7.5.3 below, condition (7.5.14) can hold as well for *nonprojectable* Lie group actions.

Remark 5

The particular cases in (7.5.13) and (7.5.17) of the mapping δ in (7.4.46) illustrate clearly why, in general, the *range* of δ has to be a union of *cartesian products* of nowhere dense differential algebras $A_{\Delta'}$, see also (7.4.45), rather than one single such an algebra, as for instance, was initially attempted in (7.3.15) and (7.4.1).

Indeed, even in the *classical* case of a *nonprojectable* Lie group action on one single $\mathcal{C}^\infty$-smooth function, as for instance in (7.5.1) - (7.5.5), the issue still comes down to the fact that, typically, the mapping α in (7.5.8) may *fail* to be injective. Therefore, as seen in (A.20), (A.21), the *best* we may obtain under such circumstances is the *family* of commutative diagrams of $\mathcal{C}^\infty$-smooth functions (7.5.10), a family which depends on $\Delta'' \in \mathcal{D}_\nu = \mathcal{D}_\alpha$.

And it is only in the special case of (7.5.14), (7.5.15), when - upon convenience - the family (7.5.13) can be simplified by being replaced with its largest member, corresponding to (7.5.16). In such a case, just as in (7.5.17), the *range* of δ will be *one single* nowhere dense differential algebra. In this special case, however, we have

$$g\,\psi = \varphi$$

that is, the Lie group action $g\,\psi$ happens to give again a $\mathcal{C}^\infty$-smooth function φ. As mentioned earlier, this always happens in the case of *projectable* Lie group actions. However, it can also happen in other more general, *nonprojectable* cases of actions, provided that the corresponding mappings α in (7.5.8), (7.5.14) turn out to be $\mathcal{C}^\infty$-smooth diffeomorphisms.

5.2. Arbitrary Lie Group Actions on Generalised Functions

Let us now turn to the *second test* case of the commutative diagram (7.4.66), which this time involves arbitrary nonlinear Lie group actions on a wide class of *generalised solutions* of, among others, arbitrary analytic nonlinear PDEs.

As mentioned, and also seen below in (7.5.26) - (7.5.29), these generalised solutions belong to the *nowhere dense differential algebras* (1.6), (7.1.4). And a special feature of these generalised solutions is related to their possible *singularities*. Namely, these singularities can be those of arbitrary analytic functions of several complex variables, and as such, near to them, the functions can exhibit a rather arbitrary behaviour, be it fast convergence to infinity, or oscillations. As a consequence, typically, these generalised solutions may *no longer* belong to the vector space of the Schwartz distributions, *nor* to the algebra of generalised functions of Colombeau, see for details Rosinger [7-9], Rosinger & Walus [1,2].

Global Cauchy - Kovalevskaia Theorem. Let us suppose that we deal with the following general *nonlinear analytic* case of the operator $T(x, D)$ in (1.1) and (1.4), when

$$(7.5.18) \qquad T(x,D)U(x) \; = \; D_t^m U(t,y) - G(t,y,\ldots,D_t^p D_y^q U(t,y),\ldots)$$

where $U : \Omega \to \mathbf{C}$ is the unknown function, while $x = (t,y) \in \Omega$, $t \in \mathbf{R}$, $y \in \mathbf{R}^{n-1}$, $p \in \mathbf{N}$, $0 \leq p < m$, $q \in \mathbf{N}^{n-1}$, $p + |q| \leq m$, and G is analytic in all of its variables.

Our interest is in finding *global generalised solutions*, that is, *on the whole of* Ω, for the nonlinear analytic PDEs

$$(7.5.19) \qquad T(x,D)U(x) \; = \; 0, \quad x \in \Omega$$

with $T(x, D)$ given in (7.5.18). For that purpose, for a given $t_0 \in \mathbf{R}$, we suppose that on the nonvoid and noncharacteristic analytic hypersurface

$$(7.5.20) \qquad S \; = \; \{\, x = (t,y) \in \Omega \mid t = t_0 \,\}$$

we are given the analytic initial value problem

$$(7.5.21) \qquad D_t^p U(t_0,y) \; = \; g_p(y), \quad 0 \leq p < m, \; (t_0,y) \in S$$

Now $T(x, D)$ in (7.5.18) will obviously define a mapping

$$(7.5.22) \quad \mathcal{C}^\infty(\Omega) \;\stackrel{T(x,D)}{\longrightarrow}\; \mathcal{C}^\infty(\Omega)$$

and in view of (7.1.6) and (7.1.8), we can extend (7.5.22) to the commutative diagram

$$(7.5.23) \qquad \begin{array}{ccc} \mathcal{C}^\infty(\Omega) & \xrightarrow{\;T(x,D)\;} & \mathcal{C}^\infty(\Omega) \\ \Big\downarrow & & \Big\downarrow \\ A_\Omega & \xrightarrow[\;T\;]{} & A_\Omega \end{array}$$

where for $s = (\psi_\nu | \nu \in \mathbf{N}) \in (\mathcal{C}^\infty(\Omega))^{\mathbf{N}}$, we have

$$(7.5.24) \qquad A_\Omega \ni U = s + \mathcal{I}_{nd}(\Omega) \;\longmapsto\; T\,U = Ts + \mathcal{I}_{nd}(\Omega) \in A_\Omega$$

with

(7.5.25) $\quad Ts \ = \ (T(x, D)\psi_\nu | \nu \in \mathbf{N})$

In Rosinger [7-9], the following *global version* of the Cauchy - Kovalevskaia theorem was proved :

Theorem G C-K

The analytic nonlinear PDE in (7.5.19), with the analytic noncharacteristic initial value problem (7.5.20), (7.5.21), has *global* generalised solutions

(7.5.26) $\quad U \in A_\Omega$

defined on the whole of Ω. These solutions U are *analytic* functions

(7.5.27) $\quad \psi : \Omega \setminus \Gamma \longrightarrow \mathbf{C}$

when they are restricted to *open dense* subsets $\Omega \setminus \Gamma \subseteq \Omega$, where the subsets

(7.5.28) $\quad \Gamma \subseteq \Omega$ closed and nowehre dense

are suitably chosen. Moreover, one can choose Γ to have zero Lebesgue measure, that is

(7.5.29) $\quad \text{mes}\, \Gamma \ = \ 0$

The Class $S_{nd}(\Omega)$ of Generalised Functions. It is useful to recall here the connection between ψ and U in (7.5.27), (7.5.26) respectively, and next we present the corresponding construction in a *stronger* form than in Rosinger [7-9], since it proves to have its own interest, which is not limited to the situation in Theorem G C-K above.

Let us, therefore, assume given any

(7.5.30) $\quad \Gamma \subseteq \Omega$ closed, nowhere dense

and

(7.5.31) $\quad \psi \in \mathcal{C}^\infty(\Omega \setminus \Gamma)$

It will be convenient to describe Γ and its neighbouhood by suitable $\mathcal{C}^\infty$-smooth functions. For that purpose, let us take any real valued function $\gamma' \in \mathcal{C}^\infty(\Omega)$, such that, Kahn, we have

(7.5.32) $\quad \Gamma \ = \ \{\, x \in \Omega \mid \gamma'(x) = 0\}, \quad \gamma' \geq 0$

Further, given any open set E, such that $\Gamma \subset E \subseteq \Omega$, let us take any real valued function $\gamma'' \in \mathcal{C}^\infty(\Omega)$, such that, Kahn, we have

$$0 \leq \gamma'' \leq 1$$

(7.5.33) $$\gamma'' = 1 \text{ on } \Gamma$$

$$\gamma'' = 0 \text{ on } \Omega \setminus E$$

Now, we can define the real valued function $\gamma \in \mathcal{C}^\infty(\Omega)$ by

(7.5.34) $$\gamma = \gamma' + 1 - \gamma''$$

and obtain

$$\gamma \geq 0$$

(7.5.35) $$\Gamma = \{ x \in \Omega \mid \gamma(x) = 0\}$$

$$\gamma \geq 1 \text{ on } \Omega \setminus E$$

Let us take any *decreasing* sequence of open subsets in Ω, namely $E_0 \supseteq E_1 \supseteq \ldots \supseteq E_\nu \supseteq \ldots$, such that $\Gamma \subseteq E_\nu$, with $\nu \in \mathbf{N}$, and

(7.5.36) $$\Gamma = \bigcap_{\nu \in \mathbf{N}} \bar{E}_\nu$$

where $\bar{E}_\nu$ denotes the topological closure of E_ν. Through (7.5.32) - (7.5.34) we obtain a corresponding sequence of real valued functions $\gamma_\nu \in \mathcal{C}^\infty(\Omega)$, with $\nu \in \mathbf{N}$, each of which satisfies (7.5.35).
Further, let us take any real valued function $b \in \mathcal{C}^\infty(\mathbf{R})$, such that $0 \leq b \leq 1$, b is nonincreasing on $(-\infty, 0]$, b is nondecreasing on $[0, \infty)$, and

$$b = 0 \text{ on } [-1/2, 1/2]$$

(7.5.37)

$$b = 1 \text{ on } \mathbf{R} \setminus [-1, 1]$$

Now, for $\nu \in \mathbf{N}$, we can define $\psi_\nu : \Omega \to \mathbf{R}$, by

(7.5.38) $$\psi_\nu(x) = \left| \begin{array}{ll} b((\nu+1)\gamma_\nu(x))\psi(x) & \text{if } x \in \Omega \setminus \Gamma \\ \\ 0 \quad \text{if } x \in \Gamma \end{array} \right.$$

We note that, given $\nu \in \mathbf{N}$, $x \in \Gamma$, then $\gamma_\nu(x) = 0$, according to (7.5.35). Hence, for some $\eta > 0$, we have $(\nu+1)\gamma_\nu(y) \in [-1/2, 1/2]$, if $y \in \Omega \cap B(x, \eta)$, where $B(x, \eta)$ denotes the open ball of radius η around x. Thus (7.5.37), (7.5.38) give $\psi_\nu = 0$ on $\Omega \cap B(x, \eta)$.
Therefore, we obtain the following $\mathcal{C}^\infty$-smooth *regularization* of ψ

(7.5.39) $$s = (\psi_\nu \mid \nu \in \mathbf{N}) \in (\mathcal{C}^\infty(\Omega))^{\mathbf{N}}$$

which coincides with ψ on $\Omega \setminus \Gamma$ in the sense that

$$(7.5.40) \qquad \begin{aligned} & \forall\, \mu, \nu \in \mathbf{N},\ \mu \le \nu \ : \\ & \psi_\nu \ = \ \psi \text{ on } \Omega \setminus E_\mu \supseteq \Omega \setminus \bar{E}_\mu \end{aligned}$$

while in view of (7.5.36) we have the *increasing* sequence of open subsets $\Omega \setminus \bar{E}_0 \subseteq \Omega \setminus \bar{E}_1 \subseteq \ \dots\ \subseteq \Omega \setminus \bar{E}_\nu \ \dots\ \subseteq \Omega \setminus \Gamma$, such that

$$(7.5.41) \qquad \bigcup_{\nu \in \mathbf{N}} (\Omega \setminus \bar{E}_\nu) \ = \ \Omega \setminus \Gamma$$

It follows in particular that

$$(7.5.42) \qquad \begin{aligned} & \forall\, \Delta \subset\subset \Omega \setminus \Gamma,\ \Delta \text{ nonvoid, open } : \\ & \exists\, \mu \in \mathbf{N} \ : \\ & \forall\, \nu \in \mathbf{N},\ \nu \ge \mu \ : \\ & \quad \psi_\nu \ = \ \psi \text{ on } \Delta \end{aligned}$$

It is easy to see that decreasing sequences of open subsets such as in (7.5.36) can be constructed. For instance, for $\nu \in \mathbf{N}$, we can take

$$(7.5.43) \qquad E_\nu \ = \ \{\, x \in \Omega \mid d(x, \Gamma) < 1/(\nu + 1) \,\}$$

where d denotes the usual Euclidean distance.

Based on (7.5.39), let us now define the generalised function

$$(7.5.44) \qquad U \ = \ s + \mathcal{I}_{nd}(\Omega) \in A_\Omega$$

then in the sense of (7.5.40), (7.5.42), we obtain

$$(7.5.45) \qquad U \ = \ \psi \text{ on } \Omega \setminus \Gamma$$

In view of the above, we consider the class of generalised functions $U \in A_\Omega$ constructed in (7.5.30) - (7.5.45), and denote it by

$$(7.5.46) \qquad S_{nd}(\Omega)$$

According to (7.1.1), (7.5.32) - (7.5.42), it is obvious that, see (7.5.30), (7.5.44), (7.5.45), the mapping

$$(7.5.47) \qquad (\Gamma, \psi) \longmapsto U \in S_{nd}(\Omega)$$

is well defined and surjective. Also, it does not depend on the auxilliary entities in (7.5.32) - (7.5.39), which were used for its construction.

This class $S_{nd}(\Omega)$ clearly contains the global generalised solutions of arbitrary nonlinear PDEs given by Theorem G C-K above, see Rosinger [7-9]. It should be noted that this class is *not* contained in the space $\mathcal{D}'(\Omega)$ of the Schwartz distributions, because of at least two reasons. First, the closed nowhere dense subsets Γ in (7.5.30) can have *arbitrarily large* positive Lebesgue measures, Oxtoby. And in case Ω has finite Lebesgue measure, the measure of Γ can be arbitrarily near to that of Ω. Second, analytic - let alone, $\mathcal{C}^\infty$-smooth - functions $\psi \in \mathcal{C}^\infty(\Omega \setminus \Gamma)$ in (7.5.31), can grow arbitrarily fast, or can oscillate arbitrarily, in the neighbourhood of Γ.

The same failure to contain $\mathcal{S}_{nd}(\Omega)$ applies to the Colombeau algebra $\mathcal{G}(\Omega)$, since in its definition polynomial type growth conditions are required, Colombeau.

In this way, we have the relations

$$\text{(7.5.48)} \qquad \begin{array}{rcl} \mathcal{D}'(\Omega) & \subset & A_\Omega \\ \mathcal{S}_{nd}(\Omega) & \subsetneq & A_\Omega \\ \mathcal{S}_{nd}(\Omega) & \not\subseteq & \mathcal{D}'(\Omega) \cup \mathcal{G}(\Omega) \end{array}$$

The Second Test. Now as a second test of the arbitrary nonlinear Lie group actions defined by the commutative diagrams (7.4.66), we shall apply them to the generalised functions

$$\text{(7.5.49)} \qquad U \in \mathcal{S}_{nd}(\Omega)$$

Accordingly, we have the relations (7.5.30) - (7.5.45), hence (7.3.16) gives

$$\text{(7.5.50)} \qquad \beta_*(U) = \text{class}_{nd} s_* \in (\mathcal{C}^\infty(\Omega, M))^{\mathbf{N}} / \approx_{nd}$$

where, see (7.2.2)

$$\text{(7.5.51)} \qquad \begin{array}{l} s_* = (\psi_{\nu *} | \nu \in \mathbf{N}) \in (\mathcal{C}^\infty(\Omega, M))^{\mathbf{N}} \\ \Omega \ni x \longmapsto \psi_{\nu *}(x) = (x, \psi_\nu(x)) \in M, \quad \nu \in \mathbf{N} \end{array}$$

Given now any group element $g \in G$, we obtain through (7.3.11)

$$\text{(7.5.52)} \qquad g\ \beta_*(U) = \text{class}_{nd} g\ s_* \in (\mathcal{C}^\infty(\Omega, M))^{\mathbf{N}} / \approx_{nd}$$

where, see (7.2.20)

$$\text{(7.5.53)} \qquad \begin{array}{l} g\ s_* = (g\ \psi_{\nu *} | \nu \in \mathbf{N}) \\ \Omega \ni x \longmapsto (g\ \psi_{\nu *})(x) = (g_1(x, \psi_\nu(x)), g_2(x, \psi_\nu(x))) \in M, \quad \nu \in \mathbf{N} \end{array}$$

We note that for the *increasing* sequence of open subsets in Ω

$$\text{(7.5.54)} \qquad \Omega_\nu = \Omega \setminus \bar{E}_\nu, \quad \nu \in \mathbf{N}$$

we have in view of (7.5.36), (7.5.41)

$$\bigcup_{\nu \in \mathbf{N}} \Omega_\nu = \Omega \setminus \Gamma$$

while the relations (7.5.40) - (7.5.42) imply

$$(7.5.55) \quad \begin{aligned} &\forall \, \nu \in \mathbf{N} : \\ &\beta_*(U|_{\Omega_\nu}) \; = \; \text{class}_{nd} u(\psi_*|_{\Omega_\nu}) \in (\mathcal{C}^\infty(\Omega_\nu, M))^{\mathbf{N}} / \approx_{nd} \end{aligned}$$

and therefore as well, see (7.2.20)

$$(7.5.56) \quad \begin{aligned} &\forall \, \nu \in \mathbf{N}, \; g \in G : \\ &g \, \beta_*(U|_{\Omega_\nu}) \; = \; \text{class}_{nd} u(g \; \psi_*|_{\Omega_\nu}) \in (\mathcal{C}^\infty(\Omega_\nu, M))^{\mathbf{N}} / \approx_{nd} \end{aligned}$$

It follows that, on each nonvoid open subset $\Omega_\nu \subset \Omega \setminus \Gamma$, with $\nu \in \mathbf{N}$, we are back to the situation in subsection 7.5.1, and then (7.5.6) will give

$$(7.5.57) \quad \begin{aligned} &\forall \, \nu \in \mathbf{N}, \; g \in G : \\ &g \, \beta_*(U|_{\Omega_\nu}) \in B_{\Omega_\nu} \subseteq B_{[\Omega]}, \quad \text{if } \Delta_{(g, \psi|_{\Omega_\nu})} \neq \phi \end{aligned}$$

We show now the *important* result that, see (7.5.6)

$$(7.5.58) \quad g \, \beta_*(U) \in B_\Omega \subseteq B_{[\Omega]}, \quad \text{if } \Omega_{(g,\psi)} \neq \phi$$

where we denoted

$$(7.5.59) \quad \Omega_{(g,\psi)} \; = \; \{ \, x \in \Omega \setminus \Gamma \mid J_\alpha(x) \neq 0 \, \}$$

where, see (2.11), (7.5.8), the mapping α is given by

$$(7.5.60) \quad \Omega \setminus \Gamma \overset{\alpha}{\longmapsto} g_1(x, \psi(x)) \in \Omega$$

It follows that

$$(7.5.61) \quad \Omega_{(g,\psi)} \overset{\alpha}{\longmapsto} \alpha(\Omega_{(g,\psi)}) \subseteq \Omega$$

is a $\mathcal{C}^\infty$-smooth surjective local diffeomorphism.

Remark 6

Property (7.5.58) clearly is a *strengthened* version of (7.5.57), as well as of (7.5.6), and it shows that B_Ω, therefore even more so $B_{[\Omega]}$, can *accommodate singularities* such as those of any $U \in S_{nd}(\Omega)$, that is, singularities given by arbitrary closed, nowhere dense subsets $\Gamma \subseteq \Omega$, of arbitrary large Lebesgue measure.

□

In order to prove (7.5.58), we first show that, see (7.5.54), (7.5.59)

$$(7.5.62) \quad \begin{aligned} &\exists \, \mu \in \mathbf{N} : \\ &\Omega_\mu \bigcap \Omega_{(g,\psi)} \neq \phi \end{aligned}$$

Indeed, if it is false, then (7.5.36), (7.5.54) imply $\Omega_{(g,\psi)} \subseteq \Gamma$, which, in view of the fact that Γ is nowhere dense, contradicts the assumption in (7.5.58) that the open subset $\Omega_{(g,\psi)}$ is nonvoid.

Let us now apply to $g\ \beta_*(U)$ the sufficient conditions at pct.2) in Remark 3. For that, we choose a fixed $\mu \in \mathbf{N}$ which satisfies (7.5.62). Further, in view of (7.5.53), let us take $(V_\nu | \nu \in \mathbf{N}) \in (\mathcal{C}^\infty(\Omega, M))^{\mathbf{N}}$, given by

$$V_\nu = g\ \psi_{\nu *}, \quad \nu \in \mathbf{N}$$

then clearly

(7.5.63) $\qquad g\ \beta_*(U) \stackrel{1}{\longmapsto} (V_\nu | \nu \in \mathbf{N})$

Now let us take, see (7.5.58), any nonvoid open subset

(7.5.64) $\qquad \Delta'' \subseteq \Omega_\mu \cap \Omega_{(g,\psi)}$

such that, see (7.5.58) - (7.5.61)

(7.5.65) $\qquad \Delta'' \stackrel{\alpha}{\longrightarrow} \alpha(\Delta'') = \Delta' \subseteq \Omega$

is a $\mathcal{C}^\infty$-smooth diffeomorphism. Then clearly,

(7.5.66) $\qquad \Delta' \subseteq \Omega \quad$ nonvoid, open

and we have the commutative diagram

(7.5.67)

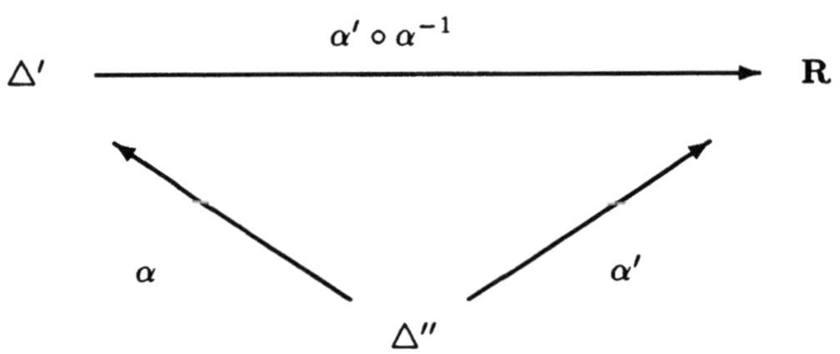

where we have defined, see (7.5.8)

(7.5.68) $\qquad \Omega \setminus \Gamma \ni x \stackrel{\alpha'}{\longmapsto} \alpha'(x) = g_2(x, \psi(x)) \in \mathbf{R}$

But (7.5.65), (7.5.40), (7.5.63) imply, see (7.5.7)

$$\forall\ \nu \in \mathbf{N},\ \nu \geq \mu :$$

$$V_\nu = g\ \psi_* \quad \text{on } \Delta''$$

therefore (7.5.63) - (7.5.67) give

$$g\ \beta_*(U) \overset{1}{\longmapsto} (V_\nu | \nu \in \mathbf{N}) \overset{2}{\longmapsto} \Delta' \overset{3}{\longmapsto} U'$$

where we have defined

$$U' = \text{class}_{nd} u(\alpha' \circ \alpha^{-1}) \in A_{\Delta'}$$

In this way $g\ \beta_*(U)$ satisfies (7.4.39), and the proof of (7.5.58) is completed.

A consequence of (7.5.58), see also (7.5.13), is that for any given $\mu \in \mathbf{N}$, we obtain

(7.5.69) $$\bigcup_{\substack{\nu \in \mathbf{N} \\ \nu \geq \mu}} \delta(g\ \beta_*(U|_{\Omega_\nu})) \subseteq \delta(g\ \beta_*(U)), \quad \text{if } \Delta_{(g,\psi|_{\Omega_\mu})} \neq \phi$$

where as before, the inclusion $\subseteq$ means that every family in the left hand term is a *subfamily* of the family in the right hand term, see (7.5.52), (7.5.53). Indeed, (7.5.69) follows easily by noting that, in view of (5.8), (7.5.54), the sequence of open subsets in Ω

$$\Delta_{(g,\psi|_{\Omega_\nu})}, \quad \nu \in \mathbf{N}$$

is *non decreasing.*

A General Method for Lie Group Actions on $S_{nd}(\Omega)$. With certain variations, the general method in (7.5.30) - (7.5.69) for constructing arbitrary nonlinear, including nonprojectable, Lie group actions on generalised functions in $S_{nd}(\Omega)$, will be used in the sequel.

The inclusion (7.5.69) suggests us to formulate

Problem 1

How *large* and *what* is the difference between the left and right hand terms in (7.5.69) ?
More precisely, are there nonvoid open subsets $\Omega' \subseteq \Omega$, such that

$$\Omega' \cap \Gamma \neq \phi$$

$$g\ \beta_*(U|_{\Omega'}) \in B_{\Omega'} \subseteq B_{[\Omega']} \quad \text{if } \Delta_{(g,U|_{\Omega'})} \neq \phi$$

and could we in such a case also compute

$$\delta(g\ \beta_*(U|_{\Omega'})) \ ?$$

Remark 7

1) It is important to point out the following *two basic steps* involved in what we are dealing with here, in subsection 7.5.2.
The *first step* was already achieved completely in (7.5.58). Let us recapitulate it here, namely

$$\forall\ U \in S_{nd}(\Omega),\ g \in G\ :$$

$$g\ \beta_*(U) \in B_\Omega \subseteq B_{[\Omega]}, \quad \text{if}\ \ \Omega_{(g,U)} \neq \phi$$

where G can be *any* Lie group. In other words, we have been able to *define arbitrary* nonlinear, including nonprojectable, Lie group actions $g\ \beta_*(U)$ on all the generalised functions $U \in S_{nd}(\Omega)$.
The *second step*, namely, *to compute* such Lie group actions, that is, to obtain $\delta(g\ \beta_*(U))$, has in part been achieved in (7.5.69), by noting that the families of generalised functions in the left hand term of this relation can, in view of (7.5.57), be computed in the simple manner of (7.5.6) in subsection 7.5.1.
However, as pointed out in Problem 1, this is only a partial solution.

2) We can note that the first step means, see (7.5.30), (7.5.31), to take a function $\psi \in \mathcal{C}^\infty(\Omega \setminus \Gamma)$ with *singularity* on the closed nowhere dense subset $\Gamma \subseteq \Omega$, then regularize it, see (7.5.44), (7.5.45), through $U \in A_\Omega$ over the whole of Ω, further, see (7.5.52), define the arbitrary Lie group action $g\ \beta_*(U)$ on the parametric representation U_* of U, and finally, see (7.5.58), show that $g\ \beta_*(U)$ is in the domain of the mapping δ, more precisely $g\ \beta_*(U) \in B_\Omega \subseteq B_{[\Omega]}$, when $\Omega_{(g,\psi)} \neq \phi$.

And as mentioned, and also seen in (7.5.58), this first step is fully accomplished under the most general conditions, although - unlike the second step - this is the step which alone comprises all the construction related to arbitrary Lie group actions on generalised functions in $S_{nd}(\Omega)$.

The second step means, see (7.4.46), to *extract* from $g\ \beta_*(U)$, through the mapping δ, all the possible generalised functions $U' \in A_{\Delta'}$, with nonvoid, open $\Delta' \subseteq \Omega$. These generalised functions $U' \in A_{\Delta'}$, are supposed to be the result of the Lie group action on the generalised function $U \in S_{nd}(\Omega) \subseteq A_\Omega$.
As mentioned, there may be an interest in those U' for which Δ' is *maximal*. However, there are in general *two* impediments which may prevent Δ' from also having the maximum size, that is, $\Delta' = \Omega$.
First, as seen in subsection 7.5.1, even if $\Gamma = \phi$, we may nevertheless have $\Delta' \neq \Omega$, owing to the simple fact that the corresponding mapping α, see (2.11), (7.5.7), may fail to be a $\mathcal{C}^\infty$-smooth diffeomorphism on the whole of Ω.
Second, since in general $\Gamma \neq \phi$, the singularity of ψ may further limit the sizes of the possible subsets Δ'.

3) Owing to the *arbitrary* nature of Γ and ψ in (7.5.30) and (7.5.31), as well as to the equally arbitrary nonlinear, and possibly nonprojectable, Lie group actions under consideration, it may appear that a general answer to Problem 1 would not be so easy. However, in the next two examples, we shall give some answers to

this problem, when it is considered in certain particular cases, and the way these answers are obtained may contribute towards a clarification of a more general interest.

In particular, Example 5 will show one of the most simple and basic difficulties involved in the second step mentioned above. This difficulty appears even in the case of a generalised function in $S_{nd}(\mathbf{R})$ as simple as the one dimensional Heaviside function H. Indeed, this happens when H is subjected to a very simple *nonprojectable* Lie group action. Then it turns out that the *only* way to represent *globally* the result of this Lie group action on H, that is, to represent it on the whole of $\mathbf{R}$ - which is the domain of definition of H, when it is seen as a generalised function - is to do so by its *parametric* representation $g\ \beta_* H$, see (7.5.94).

5.3. Two Examples

Here we present two examples which in chapter 8 will help us when we apply the nonprojectable Lie group symmetries of the nonlinear shock wave equation to its rarefaction waves, and to certain instances of its Riemann solvers.

Both of these examples involve generalised functions from $S_{nd}(\Omega)$, and therefore, the respective nonprojectable Lie group actions on them will be constructed following the ideas of the general method in (7.5.30) - (7.5.69).

Example 4

Let us consider the one dimensional, $n = 1$, case of (7.5.30), (7.5.31), when

$$(7.5.70) \qquad \psi = x_+$$

that is, $\psi(x) = 0$, for $x \leq 0$, and $\psi(x) = x$, for $x \geq 0$. In this case,

$$(7.5.71) \qquad \Omega = \mathbf{R}, \quad \Gamma = \{0\}, \quad \psi \in \mathcal{C}^\infty(\Omega \setminus \Gamma)$$

which in chapter 8 will help us in constructing the Lie group symmetries of the rarefaction waves.

Our aim here in this example is to compute, see (7.5.58), (7.5.69), (7.4.46)

$$(7.5.72) \qquad \delta(g\ \beta_*(x_+)) = (\ U_i' \mid i \in I\ ) \quad \text{when } \Omega_{(g,x_+)} = \mathbf{R}_{(g,x_+)} \neq \phi$$

for certain *nonprojectable* Lie group actions $g \in G$. Here we have, see (7.4.42)

$$(7.5.73) \qquad g\ \beta_*(x_+) \overset{5}{\longmapsto} (\ (\ \Delta_i',\ U_i'\ ) \mid i \in I\ )\ )$$

where $\Delta_i' \subseteq \mathbf{R}$ is nonvoid, open, while $U_i' \in A_{\Delta_i'}$, for $i \in I$, and as before, we shall be interested in the *largest*, that is, *maximal* nonvoid open subsets Δ_i'.

For that purpose, we follow the relevant part of the earlier construction in (7.5.32) - (7.5.69), with some simplifications which are allowed by our particular function in (7.5.70). We start by taking a function $b \in \mathcal{C}^\infty(\mathbf{R})$, such that

$$\text{supp}\ b = [0,1]$$

(7.5.74) $$\int_{\mathbf{R}} b(x)dx = 1$$

$$0 \leq b \leq K$$

where $K > 1$. Further, we define the sequence of functions $c_\nu \in \mathcal{C}^\infty(\mathbf{R})$, with $\nu \in \mathbf{N}$, by

(7.5.75) $$c_\nu(x) = \int_{-\infty}^{(\nu+1)x} b(y)dy, \quad x \in \mathbf{R}$$

Finally, we give the generalised function $U = x_+ \in S_{nd}(\mathbf{R})$ a representation similar to (7.5.44), (7.5.38), (7.5.39), and for that purpose, for $\nu \in \mathbf{N}$, we define the $\mathcal{C}^\infty$-smooth function

(7.5.76) $$\psi_\nu(x) = \left| \begin{array}{l} c_\nu(x)x \ \text{ if } x > 0 \\ \\ 0 \ \text{ if } x \leq 0 \end{array} \right.$$

hence we have

$$\psi_\nu = \psi = x_+ \ \text{ on } (-\infty, 0] \cup [1/(\nu+1), \infty)$$

(7.5.77) $$0 \leq \psi_\nu \leq x_+ \ \text{ on } \mathbf{R}$$

$$0 \leq \psi_\nu' \leq K+1 \ \text{ on } \mathbf{R}$$

Let us now consider given an arbitrary Lie group G acting on $M = \Omega \times \mathbf{R} = \mathbf{R}^2$ as in (1.2). Then for a fixed $g \in G$, with the notation in (7.5.39), (7.5.44), (7.5.50) - (7.5.53), we obtain

(7.5.78) $$g(x_+)_* = \text{class}_{nd}(g\psi_{\nu *} \mid \nu \in \mathbf{N}) \in (\mathcal{C}^\infty(\mathbf{R}, \mathbf{R}^2))^{\mathbf{N}} / \approx_{nd}$$

If for $\nu \in \mathbf{N}$, we denote, see (7.5.63)

(7.5.79) $$V_\nu = g\psi_{\nu *} \in \mathcal{C}^\infty(\mathbf{R}, \mathbf{R}^2)$$

then (7.4.19) gives the $\mathcal{C}^\infty$-smooth surjective local diffepmorphism

(7.5.80) $$D_\nu \xrightarrow{V_{\nu 1}} \Delta_\nu' = V_{\nu 1}(D_\nu)$$

where, see (7.4.18)

(7.5.81) $$D_\nu = \{\, x \in \mathbf{R} \mid D_x\, g_1(x, \psi_\nu(x)) \neq 0 \,\}$$

Also, for $\nu \in \mathbf{N}$, $x \in \mathbf{R}$, we have

(7.5.82) $$D_x\, g_1(x, \psi_\nu(x)) = g_{1x}(x, \psi_\nu(x)) + g_{1u}(x, \psi_\nu(x))\psi_\nu'(x)$$

And now, in order to be more specific, and similar to the Lie group in Example 1, see also (6.9), let us consider the *nonprojectable* Lie group action

$$(7.5.83) \qquad \begin{aligned} \bar{x} &= x + \epsilon u \\ \bar{u} &= u \end{aligned}$$

where $x, u \in \mathbf{R}$, $\epsilon = g \in G = (\mathbf{R}, +)$. Then (7.5.82) becomes

$$D_x \; g_1(x, \psi_\nu(x)) = 1 + \epsilon\psi'_\nu(x), \quad \nu \in \mathbf{N}$$

Now (7.5.81) gives for $\nu \in \mathbf{N}$

$$(7.5.84) \qquad D_\nu \; = \; \{ \; x \in \mathbf{R} \mid 1 + \epsilon\psi'_\nu(x) \; \neq \; 0 \; \}$$

thus (7.5.76), (7.5.77) imply

$$(7.5.85) \qquad \begin{aligned} &D_\nu \; = \; \mathbf{R} \quad \text{if } \epsilon > -1/(K+1) \\ &D_\nu \supseteq (-\infty, 0] \, \cup \, [1/(\nu+1), \infty) \quad \text{if } \epsilon \neq -1 \\ &D_\nu \supseteq (-\infty, 0] \quad \text{if } \epsilon \; = \; -1 \end{aligned}$$

In view of (7.5.85), and for convenience, we shall limit ourselves to the case when

$$(7.5.86) \qquad -1/(K+1) \; < \; \epsilon \; = \; g \; < \; \infty$$

Then, for $\nu \in \mathbf{N}$, we have the $\mathcal{C}^\infty$-smooth *diffeomorphisms*

$$(7.5.87) \qquad \Omega \; = \; \mathbf{R} \; = \; D_\nu \xrightarrow{V_{\nu 1}} \Delta'_\nu \; = \; V_{\nu 1}(\mathbf{R}) \; = \; \mathbf{R}, \qquad \nu \in \mathbf{N}$$

and in view of (5.8), we also have

$$(7.5.88) \qquad \Omega_{(g, x_+)} \; = \; \mathbf{R}_{(\epsilon, x_+)} \; = \; \mathbf{R} \; = \; \Omega \neq \phi$$

In this way, we obtain, for $\nu \in \mathbf{N}$, the following version of the commutative diagrams in (7.4.32)

(7.5.89)

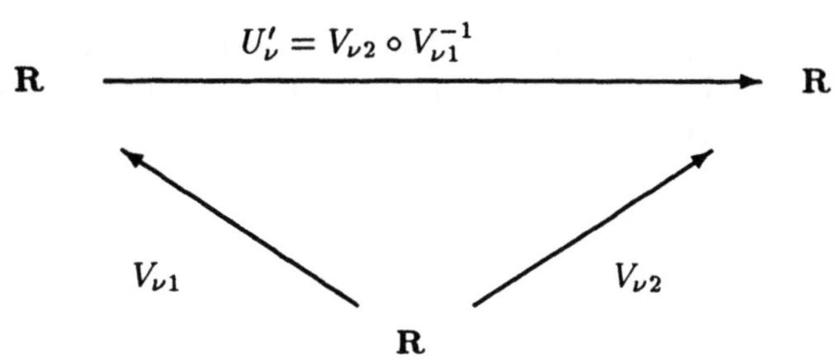

Let us now recall that (7.5.58), (7.5.88) give

$$(7.5.90) \qquad g \; \beta_*(x_+) \in B_{\mathbf{R}} \subseteq B_{[\mathbf{R}]}$$

Further, in view of (7.5.89), the conditions (7.4.59) - (7.4.62) from pct.4) in Remark 4 are satisfied. Therefore, (7.4.63) gives, see also the comment related to (7.4.58)

$$(7.5.91)\quad \begin{aligned}\delta(g\ \beta_*(x_+)) &= U' = \\ &= \text{class}_{nd}(U'_\nu = V_{\nu 2}\circ V_{\nu 1}^{-1}\ \mid \nu\in\mathbf{N})\in A_{\mathbf{R}}\end{aligned}$$

hence we ***fully recover*** the classical, simplified form of the mapping in (7.4.51). Moreover, (7.5.91) gives a particular answer to Problem 1, as this clearly follows from (7.5.90).

Example 5

Here we deal with an example which in chapter 8 will help us with the nonprojectable Lie group symmetries of the Riemann solvers of the nonlinear shock wave eqution.
For that, let us consider certain nonprojectable Lie group actions on the Heaviside function in one dimension, $n = 1$. Let therefore

$$(7.5.92)\qquad \chi = H$$

hence $\chi(x) = 0$, for $x < 0$, and $\chi(x) = 1$, for $x > 0$. Similar with Example 4, we then have

$$(7.5.93)\quad \Omega = \mathbf{R},\ \ \Gamma = \{0\},\ \ \chi\in\mathcal{C}^\infty(\Omega\setminus\Gamma)$$

As is known, H is the *distributional* derivative of x_+ in (7.5.70), moreover, in the classical sense

$$(7.5.94)\qquad (x_+)' = H \quad \text{on } \mathbf{R}\setminus\{0\}$$

Here again, we shall be interested to compute, see (7.5.72), (7.5.73)

$$(7.5.95)\quad \delta(g\ \beta_* H) = (\ U'_i \mid i\in I\ ) \quad\text{when}\quad \Omega_{(g,H)} = \mathbf{R}_{(g,H)} \neq \phi$$

for the simple nonprojectable Lie group actions in (7.5.83).

We shall start by giving $U = H\in S_{nd}(\mathbf{R})$ a representation similar to (7.5.44), (7.5.38), (7.5.39), or (7.5.76). For that purpose, let us take the sequence of $\mathcal{C}^\infty$-smooth functions, see (7.5.75)

$$(7.5.96)\qquad \chi_\nu = c_\nu,\quad \nu\in\mathbf{N}$$

Now, for a given $g\in G$, similar with (7.5.78), we obtain

$$(7.5.97)\quad g(H)_* = \text{class}_{nd}(g\ \chi_{\nu *}\mid \nu\in\mathbf{N})\in(\mathcal{C}^\infty(\mathbf{R},\mathbf{R}^2))^{\mathbf{N}}/\approx_{nd}$$

Then for $\nu\in\mathbf{N}$, we denote this time, see (7.5.63)

$$(7.5.98)\quad V_\nu = g\ \chi_{\nu\ *}\in\mathcal{C}^\infty(\mathbf{R},\mathbf{R}^2)$$

It follows that for the nonprojectable Lie group action (7.5.83), we have for $\nu\in\mathbf{N}$, see (7.4.18)

(7.5.99) $\quad D_\nu = \{ x \in \mathbf{R} \mid 1 + \epsilon c'_\nu(x) \neq 0 \}$

therefore, see (7.5.74), (7.5.75)

$$(7.5.100) \qquad \begin{aligned} D_\nu &= \mathbf{R} \quad \text{if } \epsilon > -1/(K(\nu+1)) \\ D_\nu &\supseteq (-\infty, 0] \cup [1/(\nu+1), \infty) \quad \text{if } \epsilon \leq -1/(K(\nu+1)) \end{aligned}$$

In view of (7.5.100) we consider the following two cases of Lie group actions in (7.5.83), namely, when

(7.5.101) $\qquad \epsilon = g \geq 0$

and when

(7.5.102) $\qquad \epsilon = g < 0$

Clearly, for (7.5.101), relations similar to (7.5.87) - (7.5.91) will hold, in particular, we obtain the corresponding versions of (7.5.87) and (7.5.89), as well as

$$(7.5.103) \qquad \begin{aligned} &\mathbf{R}_{(g,H)} \neq \phi \\ &g\ \beta_* H \in B_{\mathbf{R}} \subseteq B_{[\mathbf{R}]} \end{aligned}$$

hence, in the sense mentioned in the comment related to (7.4.58), we have

$$(7.5.104) \qquad \begin{aligned} \delta(g\ \beta_* H) &= U' = \\ &= \text{class}_{nd}(\ U'_\nu = V_{\nu 2} \circ V_{\nu 1}^{-1} \mid \nu \in \mathbf{N}\ ) \in A_{\mathbf{R}} \end{aligned}$$

Let us now turn to the less simple case of (7.5.102). Given $\epsilon < 0$, let $\mu \in \mathbf{N}$ be such that $\epsilon \leq -1/(\mu+1)$, then, see (7.5.74)

$$\epsilon \leq -1/(\nu+1) \leq -1/(K(\nu+1)), \quad \nu \in \mathbf{N},\ \nu \geq \mu$$

and thus, see (7.5.100)

(7.5.105) $\quad D_\nu \supseteq (-\infty, 0] \cup [1/(\nu+1), \infty), \quad \nu \in \mathbf{N},\ \nu \geq \mu$

We recall that, according to (7.4.19), we have for $\nu \in \mathbf{N}$ the $\mathcal{C}^\infty$-smooth local surjective diffeomorphisms

$$D_\nu \xrightarrow{V_{\nu 1}} V_{\nu 1}(D_\nu) \subseteq \Omega$$

however, as in (7.5.72), (7.5.73), see also (7.4.32) and pct.2) in Remark 3, we are interested in subsets $\Delta''_\nu \in \mathcal{D}_\nu$ which, in view of (7.4.22), will result in $\mathcal{C}^\infty$-smooth diffeomorphisms

(7.5.106) $\quad \Delta''_\nu \xrightarrow{V_{\nu 1}} V_{\nu 1}(\Delta''_\nu) = \Delta' \subseteq \Omega$

with the largest possible nonvoid open subsets Δ'. Let us therefore denote for $\nu \in \mathbf{N}$, see (7.5.105)

$$\begin{aligned} \Delta''_1 &= (-\infty, 0) \\ \Delta''_{2\,\nu} &= (1/(\nu+1), \infty) \end{aligned} \qquad (7.5.107)$$

then (7.5.98), (7.5.83) imply

$$V_{\nu 1}(x) = \begin{cases} x & \text{if } x \in \Delta''_1 \\ x + \epsilon & \text{if } x \in \Delta''_{2\,\nu} \end{cases} \qquad (7.5.108)$$

hence, for $\nu \geq \mu$, we have the $\mathcal{C}^\infty$-smooth diffeomorphisms

$$\begin{aligned} \Delta''_1 &\xmapsto{V_{\nu 1}} \Delta'_1 = (-\infty, 0) \\ \Delta''_{2\,\mu} &\xmapsto{V_{\nu 1}} \Delta'_{2\,\mu} = (1/(\nu+1)+\epsilon, \infty) \end{aligned} \qquad (7.5.109)$$

and for each of them, we obtain the corresponding versions of the commutative diagrams (7.4.32)

The first sequence of mappings in (7.5.109) and the corresponding commutative diagrams (7.4.32) give the generalised function

$$U'_1 = \text{class}_{nd}(\, U'_{1\,\nu} = V_{\nu 2} \circ V_{\nu 1}^{-1} \mid \nu \in \mathbf{N} \,) \in A_{(-\infty,0)} \qquad (7.5.110)$$

With respect to the second sequence of mappings in (7.5.109) and the corresponding commutative diagrams (7.4.32) we note that, in view of subsection 7.4.8., they can be extended from $\Delta''_{2\,\nu}$ to $\Delta''_2 = (0, \infty)$. In this case $\Delta'_{2\,\nu}$ in (7.5.109) becomes $\Delta'_2 = (\epsilon, \infty)$, and then, similar with (7.5.110), we obtain the generalised function

$$U'_2 = \text{class}_{nd}(\, U'_{2\,\nu} = V_{\nu 2} \circ V_{\nu 1}^{-1} \mid \nu \in \mathbf{N} \,) \in A_{(\epsilon,\infty)} \qquad (7.5.111)$$

In this way, the respective versions of (7.5.103), (7.5.104) become

$$\begin{aligned} &\mathbf{R}_{(g,H)} \neq \phi \\ &g\ \beta_* H \in B_{\mathbf{R}} \subseteq B_{[\mathbf{R}]} \\ &\delta(g\ \beta_* H) \supseteq (\, U'_1,\ U'_2 \,) \end{aligned} \qquad (7.5.112)$$

However, clearly, in the case of (7.5.102), we have

$$(-\infty, 0) \cap (\epsilon, \infty) = (\epsilon, 0) \neq \phi$$

therefore $\delta(g\ \beta_* H)$ *cannot* contain a generalised function $U' \in A_{\Delta'}$, with the largest or maximum size $\Delta' \subseteq \mathbf{R}$. In particular, we obtain that

(7.5.113) $\Delta' = \mathbf{R}$ cannot hold

which it will be important to remember in chapter 8, when we deal with the nonprojectable Lie group actions on the Riemann solvers of the nonlinear shock wave equation.

Nevertheless, the fact that, for $\epsilon = g < 0$, the generalised function

(7.5.114) $U'_2 \in A_{(\epsilon,\infty)}$

belongs to $\delta(g\ \beta_* H)$, still gives a particular solution to Problem 1, since in our case (7.5.102), we clearly have

(7.5.115) $(\epsilon,\infty) \cap \Gamma \neq \phi$

Here it is also important to note the second fact, namely that, see (7.5.110), (7.5.111), the domains of definition $(-\infty, 0)$ and (ϵ,∞) of U'_1 and U'_2 respectively, are *not* maximal. Indeed, let us take any

(7.5.116) $\epsilon + 1/(\mu+1) \leq \eta \leq 0$

and define

(7.5.117) $\Delta''_\eta = (-\infty, \eta) \cup (\eta - \epsilon, \infty)$

then for $\nu \geq \mu$, it follows that

(7.5.118) $\Delta''_\eta \overset{V_{\nu 1}}{\longmapsto} \Delta'_\eta = (-\infty, \eta) \cup (\eta, \infty)$

is also a $\mathcal{C}^\infty$-smooth diffeomorphism. Hence again, through the respective version of the commutative diagrams (7.4.32), we obtain the generalised function

(7.5.119) $U'_\eta \ = \ \text{class}_{nd}(\ U'_{\eta\ \nu} = V_{\nu 2} \circ V_{\nu 1}^{-1} \mid \nu \in \mathbf{N}\ ) \in A_{\Delta'_\eta}$

We note that for $\eta = 0$, we have

(7.5.120) $(-\infty, 0) \subset \Delta'_0$

Further, similar with the procedure used above for U'_2, here also, and regardless of (7.5.116), we can tend with η to ϵ, so that (7.5.119) will remain valid in the limit, when $\eta = \epsilon$. In this way, we obtain

(7.5.121) $(\epsilon, \infty) \subset \Delta'_\epsilon$

In view of (7.5.120), (7.5.121), it follows that the domains of definition of U'_1 and U'_2 in (7.5.110) and (7.5.111) are strictly contained in those of U'_η, for appropriate values of η.

Furthermore, the nonvoid open subsets $\Delta'_\eta \subseteq \mathbf{R}$, with $\eta \in [\epsilon, 0]$ are *maximal*, albeit *not* connected, under the condition that, see (7.5.118)

(7.5.122) $\Delta''_{\eta_\nu} \overset{V_{\nu 1}}{\longmapsto} \Delta'_{\eta_\nu} = (-\infty, \eta_\nu) \cup (\eta_\nu, \infty)$

are $\mathcal{C}^\infty$-smooth diffeomorphisms, where η_ν satisfies (7.5.116) and tends to ϵ, when $\nu \to \infty$.

However, it is important to note that, here, we face a far larger *complication* with respect to the structure of the maximal domains $\Delta' \subseteq \Omega$, as well as of the respective generalised functions $U' \in A_{\Delta'}$ which may belong to $\delta(g\ \beta_* H)$.
This complication has *two* aspects.
First, there are *many other* such maximal open subsets, in addition to Δ'_η, with $\eta \in [\epsilon, 0]$, see (7.5.118). This fact, which is also suggested by Example A in the Appendix, will be illustrated next in (7.5.123) - (7.5.129).
Second - and beyond the complications mentioned in Appendix, or at pct. 2) - 4) in Remark 4 in section 7.4 - is the fact that the mentioned maximal nonvoid open subsets $\Delta' \subseteq \Omega$ are not an end in themselves, but rather, they are merely the domains of definition of the generalised functions $U' \in A_{\Delta'}$ which are the actual objects of our main interest. And the way these generalised functions U' are defined, through the corresponding commutative diagrams (7.4.32), makes them depend also on nonvoid open subsets $\Delta''_\nu \subseteq \Omega$, such as for instance $\Delta''_\eta \subseteq \Omega$ in (7.5.117), (7.5.118), which give the generalised functions U'_η in (7.5.119).

This second aspect leads to the situation that the *maximality* of the domain Δ' may *depend* on such nonvoid open subsets $\Delta''_\nu \subseteq \Omega$, thus, also on the generalised functions U' to which they give rise through the respective versions of the commutative diagrams (7.4.32).
More precisely, when trying to simplify the mapping δ, as at pct. 2) - 4) in Remark 4, we may rather be concerned with the maximality of the generalised functions $U' \in A_{\Delta'}$ themselves, than merely with that of their domains Δ', see (7.5.134), (7.5.135) below.

Let us give some illustration of these two kind of possible complications. For the first one, let us take

(7.5.123) $\quad \epsilon + 1/(\mu + 1) \leq \eta_1 < \eta_2 \leq 0$

and define

(7.5.124) $\quad \Delta''_{\eta_1\ \eta_2} = (-\infty, \eta_1) \cup (\eta_2, 0) \cup (\eta_1 - \epsilon, \eta_2 - \epsilon) \cup (-\epsilon, \infty).$

Then as above, for $\nu \geq \mu$, it follows that

(7.5.125) $\quad \Delta''_{\eta_1\ \eta_2} \overset{V_{\nu 1}}{\longmapsto} \Delta'_{\eta_1\ \eta_2} = \mathbf{R} \setminus \{\eta_1, \eta_2, 0\}$

is also a $\mathcal{C}^\infty$-smooth diffeomorphism, and the corresponding form of the commutative diargams (7.4.32) will give the generalised function

(7.5.126) $\quad U'_{\eta_1\ \eta_2} \ = \ \text{class}_{nd}(\ U'_{\eta_1\ \eta_2\ \nu} = V_{\nu 2} \circ V_{\nu 2}^{-1} \mid \nu \in \mathbf{N}\ ) \in A_{\Delta'_{\eta_1\ \eta_2}}$

and again, we can keep (7.5.126), when $\eta_1 \to \epsilon$.

Clearly, the above procedure can be pursued for any

$$(7.5.127) \quad \epsilon \leq \eta_1 < \eta_2 < \ldots < \eta_l \leq 0$$

with arbitrary $l \in \mathbf{N}$, and we obtain

$$(7.5.128) \quad U'_{\eta_1 \ldots \eta_l} \in A_{\Delta'_{\eta_1 \ldots \eta_l}}$$

where

$$(7.5.129) \quad \Delta'_{\eta_1 \ldots \eta_l} \supseteq \mathbf{R} \setminus \{\eta_1, \ldots, \eta_l, 0\}$$

For the second complication let us take, see (7.5.116), (7.5.123)

$$(7.5.130) \quad \epsilon + 1/(\mu + 1) \leq \eta_1 \leq \eta \leq \eta_2 \leq 0$$

then (7.5.118), (7.5.119), (7.5.125), (7.5.126) yield the nonvoid open subsets

$$(7.5.131) \quad \Delta'_\eta = \mathbf{R} \setminus \{\eta\}, \quad \Delta'_{\eta_1 \eta_2} = \mathbf{R} \setminus \{\eta_1, \eta_2, 0\}$$

and the corresponding generalised functions

$$(7.5.132) \quad U'_\eta \in A_{\Delta_\eta}, \quad U'_{\eta_1 \eta_2} \in A_{\Delta_{\eta_1 \eta_2}}$$

Now we recall that, see (7.5.122), the subsets Δ'_η are maximal. Further, for

$$(7.5.133) \quad \eta = \eta_1, \ \ \eta = \eta_2 \text{ or } \eta = 0$$

in view of (7.5.131), we have

$$(7.5.134) \quad \Delta'_{\eta_1 \eta_2} \subset \Delta'_\eta$$

However, we *cannot* eliminate the generalised function $U'_{\eta_1 \eta_2}$, and replace it with U'_η, simply because of the inclusion (7.5.134), according to which the domain of the latter generalised function contains that of the former.
Indeed, it is easy to see that, as a generalised function, $U'_{\eta_1 \eta_2}$ is *not* a restriction of U'_η, since (7.5.119), (7.5.126) give

$$(7.5.135) \quad \begin{aligned} U'_\eta(x) &= \left| \begin{array}{l} x \ \text{ if } x < \eta \\ \\ x + \epsilon \ \text{ if } x > \eta \end{array} \right. \\ U'_{\eta_1 \eta_2}(x) &= \left| \begin{array}{l} x \ \text{ if } x < \eta_1 \ \text{ or } \eta_2 < x < 0 \\ \\ x + \epsilon \ \text{ if } \eta_1 < x < \eta_2 \ \text{ or } 0 < x \end{array} \right. \end{aligned}$$

Clearly, the relations (7.5.130) - (7.5.135) will significantly *complicate* the problem of the simplification of the mapping δ, as formulated initially at pct. 2) - 4) in Remark 4 in section 7.4.

Remark 8

It is *important* to note the fact in Example 5, that for $\epsilon = g < 0$ we have

$$\begin{aligned} &\delta(g\ \beta_* H) \supseteq (\ U_1',\ U_2'\ ) \\ (7.5.136) \qquad & U_1' \in A_{(-\infty,0)}, \quad U_2' \in A_{(\epsilon,\infty)} \\ &(-\infty, 0) \cap (\epsilon, \infty) \neq \phi \end{aligned}$$

and as it is easy to see, the generalised functions U_1' and U_2' are *not* equal, when restricted to $(\epsilon, 0)$, which is the intersection of their domain.
Therefore, as a *stronger* version of the negative result in (7.5.113), it follows that $\delta(g\ \beta_* H)$ *cannot* contain a largest or maximum generalised function. That is, there cannot exist $U' \in A_{\Delta'}$, with a nonvoid open subset $\Delta' \subseteq \mathbf{R}$, such that if, see (7.5.95)

$$(7.5.137) \quad \delta(g\ \beta_* H) \ = \ (\ U_i' \mid i \in I\ )$$

then

$$\begin{aligned} & \exists\ i_0 \in I\ : \\ & \quad *)\ U' = U_{i_0}' \\ (7.5.138) \qquad & \quad **)\ \forall\ i \in I,\ i \neq i_0\ : \\ & \qquad U_i' \text{ is a restrcition of } U' \end{aligned}$$

However, this is *not* so much related to the *discontinuity* of H at $0 \in \mathbf{R}$, but rather to the fact that the Lie group action (7.5.83) is *not* projectable.
Indeed, Example A in the Appendix illustrates a similar situation in the case of a function ψ given by a polynomial.

Here, in order to further help clarifying the situation, we shall consider the $\mathcal{C}^\infty$-smooth function

$$\psi(x) = \text{arctg } x, \qquad x \in \mathbf{R}$$

which, just like the Heaviside function in Example 5, is both bounded and nondecreasing on $\mathbf{R}$.
The nonprojectable Lie group action (7.5.83) will therefore give for $\epsilon = g \in G = (\mathbf{R}, +)$ the relation

$$(g\ \beta_* \psi)(x) = (x + \epsilon \text{ arctg } x, \text{arctg } x), \qquad x \in \mathbf{R}$$

hence, see (2.11), (7.5.7)

$$\alpha(x) = x + \epsilon \text{ arctg } x, \qquad x \in \mathbf{R}$$

Now, if $\epsilon < -1$, then we have

$$D_\alpha \ = \ \{\ x \in \mathbf{R} \mid \alpha'(x) \neq 0\ \} \ = \ \mathbf{R} \setminus \{-a, a\}$$

where we have denoted $a = \sqrt{-1-\epsilon}$.
Therefore, it is easy to see that we are back to the situation in Example A in the Appendix, which means that, again, $\delta(g\ \beta_* \psi)$ *cannot* contain generalised functions $U' \in A_{\Delta'}$, with the nonvoid open subset $\Delta' \subseteq \mathbf{R}$ being the largest, that is, of maximum size.
Also, relations similar to (7.5.122) will hold.

6. The Spaces $C_{[\Omega]}$ and $D_{[\Omega]}$ and Their Extensions

We can now return to the problem of the precise definition of the spaces $C_{[\Omega]}$ and $D_{[\Omega]}$ in (7.4.66) and (7.4.67), spaces which relate to the fundamental commutative diagram (7.4.65), which through (7.4.64), attempts to define the arbitrary nonlinear, including nonprojectable, Lie group actions on the generalised functions in the nowhere dense differential algebras (1.6), (1.7).

In order to do so, it will prove to be convenient to take a more general and also *simpler* view of the problem, and define the spaces

$$(7.6.1)\qquad \begin{array}{l} \mathcal{V}_\Omega \subseteq G \times ((\mathcal{C}^\infty_{par}(\Omega, M))^{\mathbf{N}} / \approx_{nd}) \\ \mathcal{W}_\Omega \subseteq G \times ((\mathcal{C}^\infty_{par}(\Omega))^{\mathbf{N}} / \approx_{nd}) \end{array}$$

in the following natural way, which will make them the largest possible, once $B_{[\Omega]}$ in (7.4.40) has been defined, namely

$$(7.6.2)\qquad \begin{array}{lcl} \mathcal{V}_\Omega & = & \{\ (g, V) \mid g\ V \in B_{[\Omega]}\ \} \\ \mathcal{W}_\Omega & = & \{\ (g, U) \mid (g, \beta_*(U)) \in \mathcal{V}_\Omega\ \} \end{array}$$

Clearly, the spaces $\mathcal{V}_\Omega$ and $\mathcal{W}_\Omega$ defined in (7.6.2) will depend on the respective Lie group G. However, as seen below, this will not cause difficulties in the applications we pursue.

The effect of this more general approach in (7.6.1), (7.6.2), is that the commutative diagram (7.4.66) becomes replaced with

$$(7.6.3)\qquad \begin{array}{ccc} \mathcal{V}_\Omega & \xrightarrow{\ \mathcal{L}_{prm}\ } & B_{[\Omega]} \\ {\scriptstyle \mathrm{id} \times \beta_*} \big\uparrow & & \big\downarrow {\scriptstyle \delta} \\ \mathcal{W}_\Omega & \xrightarrow[\ \mathcal{L}\ ]{} & A_{[\Omega]} \end{array}$$

Now, let us recall the property in (7.5.6), which means that for an arbitrary nonlinear Lie group G, if given $g \in G$ and a $\mathcal{C}^\infty$-smooth function $\psi : \Delta \to \mathbf{R}$, then provided that $\Delta_{(g,\psi)} \neq \phi$, we shall have

$$(7.6.4) \qquad (g, \text{class}_{nd} u(\psi)) \in \mathcal{W}_\Omega$$

Similarly, (7.5.58) implies that, given any $g \in G$ and a generalised function $U \in \mathcal{S}_{nd}(\Omega) \subseteq A_\Omega$, we have

$$(7.6.5) \qquad (g, U) \in \mathcal{W}_\Omega$$

whenever $\Omega_{(g,\psi)} \neq \phi$, where the connection between U and ψ is given by (7.5.45).

We can now return once more to the spaces $C_{[\Omega]}$ and $D_{[\Omega]}$ in the commutative diagram (7.4.66), and in view of (7.6.1) - (7.6.3), define

$$(7.6.6) \qquad C_{[\Omega]} = \left\{ V \in (\mathcal{C}^\infty_{par}(\Omega, M))^{\mathbf{N}} / \approx_{nd} \left| \begin{array}{l} \forall\ g \in G : \\ \quad g\ V \in B_{[\Omega]} \end{array} \right. \right\}$$

and

$$(7.6.7) \qquad D_{[\Omega]} = \{ U \in (\mathcal{C}^\infty_{par}(\Omega))^{\mathbf{N}} / \approx_{nd} \ \mid \beta_*(U) \in C_{[\Omega]} \}$$

We note that in view of (7.6.2), we have equivalently

$$(7.6.8) \qquad C_{[\Omega]} = \left\{ V \in (\mathcal{C}^\infty_{par}(\Omega, M))^{\mathbf{N}} / \approx_{nd} \left| \begin{array}{l} \forall\ g \in G : \\ \quad (g, V) \in \mathcal{V}_\Omega \end{array} \right. \right\}$$

Clearly, with the above definitions (7.6.6), (7.6.7), the spaces $C_{[\Omega]}$ and $D_{[\Omega]}$ may depend on the Lie group G considered. Nevertheless, (7.5.6) and (7.5.58) mean that given any nonlinear Lie group G, then for all $g \in G$, we shall have

$$(7.6.9) \qquad (g, \text{class}_{nd} u(\psi)),\ (g, U) \in \mathcal{W}_{[\Omega]}$$

provided that $\Delta_{(g,\psi)} \neq \phi$, respectively $\Omega_{(g,U)} \neq \phi$.
In other words, for functions $\psi \in \mathcal{C}^\infty(\Delta)$, with nonvoid open $\Delta \subseteq \Omega$, and for generalised functions $U \in \mathcal{S}_{nd}(\Omega)$, the property (7.6.9) is *universal* with respect to arbitrary nonlinear Lie groups.

It should be noted that the space $B_{[\Omega]}$ in the commutative diagram (7.6.3) does *not* depend on any Lie group, see its definition in (7.4.35).
On the other hand, owing to (7.6.9), the possible dependence of the spaces $C_{[\Omega]}$ and $D_{[\Omega]}$ on the Lie groups considered is not going to cause difficulties, when arbitrary Lie group actions are performed on usual functions in $\mathcal{C}^\infty_{par}(\Omega)$ or on generalised functions in $\mathcal{S}_{nd}(\Omega)$. In particular, we shall not have difficulties in the application to rarefaction waves and Riemann solvers of their nonprojectable Lie group symmetries, dealt with in the next chapter.

Remark 9

1) Here we can conclude that our aim stated in (7.4.64) - (7.4.67), namely, to define arbitrary nonlinear Lie group actions on generalised functions in the nowhere dense differential agebras (1.6), has been achieved, provided that we are ready to live with the restriction imposed alone by the mapping δ in (7.3.15), (7.4.46).
This restriction first leads to the restricted domain $B_{[\Omega]}$ of definition of δ. And as a consequence, as seen in this section, we are led to the further restrictions, which give us $C_{[\Omega]}$, $D_{[\Omega]}$, $\mathcal{V}_\Omega$ and $\mathcal{W}_\Omega$.

2) There is, however, an *important* point to note. Namely, as seen in (7.5.123) - (7.5.135) in Example 5, and also in Remark 8, when nonprojectable Lie groups act upon generalised functions from $\mathcal{S}_{nd}(\Omega)$, it may happen that the computation of the mapping δ proves to be too complicated from the point of view of our particular interest.
In such a case, it is more convenient *not* to use the *whole* of the commutative diagram (7.4.65), respectively (7.4.66), (7.6.3).
In other words, instead of trying to define and use the Lie group actions $\mathcal{L}$ according to (7.3.12) - (7.3.15), and correspondingly (7.4.64) - (7.4.67), that is, given by

$$(7.6.10) \qquad \mathcal{L} = \delta \circ \mathcal{L}_{prm} \circ (\mathrm{id}_G \times \beta_*)$$

we may find it more useful *not to use* the mapping δ, which is supposed to bring us back to the nowhere dense differential algebras (1.6), (1.7).
In this case (7.4.65), and therefore (7.4.64) and (7.6.10) as well, will be replaced with the commutative diagram

$$(7.6.11) \qquad \begin{array}{ccc} G \times ((\mathcal{C}^\infty_{par}(\Omega))^{\mathbf{N}} / \approx_{nd}) & \xrightarrow{\quad \mathcal{L}_* \quad} & (\mathcal{C}^\infty_{par}(\Omega, M))^{\mathbf{N}} / \approx_{nd} \\ & \mathrm{id}_G \times \beta_* \searrow \qquad \nearrow \mathcal{L}_{prm} & \\ & G \times ((\mathcal{C}^\infty_{par}(\Omega, M))^{\mathbf{N}} / \approx_{nd}) & \end{array}$$

The *advantage* in such a case is that we may perform *without any restrictions* arbitrary nonlinear, including nonprojectable, Lie group actions on all of the generalised functions in the nowhere dense differential algebras (1.6). Indeed, it is the mapping δ alone which has imposed upon us all the above restrictions, starting with that of its domain of definition $B_{[\Omega]}$, see (7.4.46), and leading to those in this section.
On the other hand, as mentioned earlier, the mapping δ is not related to any kind of Lie group action, and the restrictions which it imposes only come from the difficulties in the inversion of $\mathcal{C}^\infty$-smooth local diffeomorphisms which fail to be injective.

Therefore, in certain situations, there may indeed be a merit in avoiding altogether the use of the mapping δ, and instead, we may choose to limit ourselves to the mentioned partial use of the commutative diagram (7.4.65).
In such a case, we do no longer need to deal with *any* of the above restrictions, and therefore, we can replace (7.4.64), and consequently (7.6.10), with

$$(7.6.12) \qquad G \times ((\mathcal{C}^{\infty}_{par}(\Omega))^{\mathbf{N}} / \approx_{nd}) \xrightarrow{\mathcal{L}_*} (\mathcal{C}^{\infty}_{par}(\Omega, M))^{\mathbf{N}} / \approx_{nd}$$

that is, according to (7.6.11), with

$$(7.6.13) \qquad \mathcal{L}_* = \mathcal{L}_{prm} \circ (\mathrm{id}_G \times \beta_*)$$

Obviously, such a replacement amounts to nothing else than *starting* with any generalised function U in the nowhere dense differential algebras, that is, see (7.3.17)

$$U \in (\mathcal{C}^{\infty}_{par}(\Omega))^{\mathbf{N}} / \approx_{nd}$$

then acting upon it by an arbitrary nonlinear Lie group element $g \in G$, and finally, *stopping* at the *parametric* form of the result, which will be but a generalised function, given this time in a parametric representation

$$(7.6.14) \qquad g\ \beta_*(U) = \mathcal{L}_{prm}(g, \beta_*(U)) = \mathcal{L}_*(g, U) \in (\mathcal{C}^{\infty}_{par}(\Omega, M))^{\mathbf{N}} / \approx_{nd}$$

where $\beta_*(U)$ is the *canonical*, see (3.5), (3.22), (3.24), (7.3.16), parametric representation of the generalised function U. However, $g\ \beta_*(U)$ is typically no longer a canonical parametric representation of a generalised function, but a rather arbitrary parametric representation. This is the very fact which in the first place has led us to introduce and use the mapping δ, that is, to compute $\delta(g\ \beta_*(U))$, in order to extract from $g\ \beta_*(U)$ the usual, nonparametric form of generalised functions.

□

Forward to the Second Track, Parametric Approach. It is important to note that, as far as the study of nonlinear partial differential equations of the general form (1.1), (1.4) is concerned, the use of classical or generalised solutions which are given in the *parametric* form, such as for instance in (7.6.14), does not cause undue difficulties.
Indeed, in Theorem 1, in chapter 5, we have seen how to deal with classical solutions given in parametric form.
In chapter 9, we shall present the respective extension to generalised functions given parametrically, that is, to elements in

$$(7.6.15) \qquad (\mathcal{C}^{\infty}_{par}(\Omega, M))^{\mathbf{N}} / \approx_{nd}$$

and more generally, to elements in

$$(7.6.16) \qquad (\mathcal{C}^{\infty}_{n}(M))^{\mathbf{N}} / \approx_{nd}$$

In other words, we shall in essence return to the Lie group actions $\mathcal{L}_*$ in (7.6.11) - (7.6.13), and to their easy natural extension to the quotient spaces in (7.6.16). This will, therefore, be the ***second track*** in our two track approach mentioned at the end of chapter 3.
And in view of the significant complication we have just seen in sections 7.4 and 7.5, and we shall still see next in chapter 8, when dealing with Lie group actions on usual, nonparametric generalised functions, the development in chapter 9 will be quite different and *simpler*.

In connection with the quotient space in (7.6.16) we note that the respective equivalence relation $\approx_{nd}$ in it must in fact be an extension of that which was defined in (7.3.1) - (7.3.3), since obviously $\mathcal{C}^\infty_{par}(\Omega, M) \subseteq \mathcal{C}^\infty_n(M)$. However, this extension can be made easily, simply by replacing the nonvoid open subsets $\Delta \subseteq \Omega$ in (7.3.2), (7.3.3), with arbitrary nonvoid open subsets $\Lambda \subseteq \mathbf{R}^n$. See chapter 9 for further details.

7. Closure Property of the Mapping δ

We can now summarize the constructions in the present chapter. Our main aim has been to define the arbitrary nonlinear, including nonprojectable, Lie group actions on generalised functions in the nowehre dense differential algebras (1.6). This has been attempted by the mapping $\mathcal{L}$ in (7.4.64), as the result of the composition of the three mappings δ, $\mathcal{L}_{prm}$ and $\mathrm{id}_G \times \beta_*$ in the commutative diagram (7.4.65). And it has been *obtained* in the somewhat restricted commutative diagram (7.6.3). Indeed, owing to the mapping δ, which is a sort of inverse of β_*, and which thus does not involve any kind of Lie group action, we have run into difficulties which we could only solve by restricting the domain of δ, when compared with that initially attempted in the commutative diagram (7.4.65). This restriction, due alone to δ, has naturally implied further restrictions on the domains of $\mathcal{L}_{prm}$ and $\mathrm{id}_G \times \beta_*$, albeit as seen in the commutative diagram (7.4.65), these latter two mappings can be defined without any restrictions whatsoever.

In this way, we have ended up with the commutative diagram (7.6.3), as giving the actual definition of an appropriately restricted version of the Lie group action $\mathcal{L}$ in (7.4.64).
This restriction, nevertheless, can still include arbitrary nonlinear, thus also non-projectable, Lie group actions on large classes of generalised functions, such as for instance, those in $S_{nd}(\Omega)$, see (7.5.58).

All this, however, is but about *one single* Lie group action on a certain generalised function, that is, one single application of the mapping

$$(7.7.1) \quad \mathcal{W}_\Omega \ni (g, U) \stackrel{\mathcal{L}}{\longmapsto} \delta(g\ \beta_*(U)) \in A_{[\Omega]}$$

where the Lie group element g and the generalised function U are given.

Furthermore, as noted in Remarks 3 - 5 and 8, and as also seen in Example 5, the mapping δ appears to be somewhat strange, both with respect to the restrictions defining its domain, and the large size of its range.

Therefore, the following *closure* problem arises :

Problem 2

Given a generalised function $U \in \mathcal{A}_\Omega$ and two Lie group elements $g, g' \in G$, then :

1) To what extent we can iterate the Lie group action $\mathcal{L}$, by first computing, see (7.4.46)

$$\mathcal{L}(g, U) \;=\; \delta(g\ \beta_*(U)) \;=\; (\ U'_i \mid i \in I\ )$$

and then, for some $i \in I$, computing

$$\mathcal{L}(g', U'_i)$$

2) In case we can iterate as at pct. 1) above, to what extent we shall have satisfied the *closure* property

(7.7.2) $\quad \mathcal{L}(g', U'_i) \;=\; \mathcal{L}(g'', U)$

where, according to the group composition in G, we denoted $g'' = g'\ g$.

□

In view of the restrictions implied by the mapping δ and reflected in the commutative diagram (7.6.3) which gives the definition of the Lie group action $\mathcal{L}$, the precise formulation of the above Problem 2 is as follows.
Given a generalised function $U \in \mathcal{A}_\Omega$ and two Lie group elements $g,\ g' \in G$, let us suppose that

$$(g, U) \in \mathcal{W} \quad \text{and} \quad \mathcal{L}(g, U) \;=\; \delta(g\ \beta_*(U)) \;=\; (\ U'_i \mid i \in I\ )$$

and that for certain $i \in I$, we also have

$$(g', U'_i) \in \mathcal{W}$$

does then follow that the *closure* property holds

(7.7.3) $\quad \mathcal{L}(g', U'_i) \;=\; \mathcal{L}(g'\ g, U) \quad ?$

In the simplified cases of δ, such as at pct. 3) - 5) in Remark 4 in section 7.4, or in subsection 7.5.1, the closure property of δ does quite obviously hold, since it reduces to the classical iteration of Lie group actions.

As far as the general situation is concerned, owing to the mentioned complexities involved in the definition of the mapping δ, and in particular, of its values, we shall only give an illustration of this closure property. Namely, we shall do so in Remark 10 in chapter 8, in the case of the actions of the nonprojectable Lie group symmetries of the nonlinear shock wave equation on certain instances of its Riemann solvers.

Chapter 8

Nonprojectable Lie Group Symmetries of Rarefaction Waves and Riemann Solvers

As mentioned in chapter 6, the nonlinear shock wave equation

(8.1) $\quad U_t + UU_x = 0, \quad t \geq 0, \quad x \in \mathbf{R}$

has one of the symmetries of its classical solutions given by the nonprojectable Lie group

$$(8.2) \qquad \begin{aligned} \bar{t} &= t \\ \bar{x} &= x + \epsilon\, a(u) \\ \bar{u} &= u \end{aligned}$$

where $\epsilon = g \in G = (\mathbf{R}, +)$ is the group parameter, while $a : \mathbf{R} \rightarrow \mathbf{R}$ can be any $\mathcal{C}^\infty$-smooth function. Hence in this case, we can consider, see (1.2)

(8.3) $\quad \Omega = (0, \infty) \times \mathbf{R}, \quad M = \Omega \times \mathbf{R} = (0, \infty) \times \mathbf{R}^2, \quad \epsilon = g \in G = (\mathbf{R}, +)$

The rarefaction wave solutions of (8.1) are, see (6.6)

(8.4) $\quad U(t, x) \;=\; u_l + (x/t - u_l)_+ - (x/t - u_r)_+, \quad t > 0, \; x \in \mathbf{R}$

where $u_l < u_r$, while the Riemann solvers of (8.1) are, see (6.8)

(8.5) $\quad U(t, x) = u_l + (u_r - u_l)\; H(x - st), \quad t > 0, \quad x \in \mathbf{R}$

where $u_l > u_r$, and $s = (u_l + u_r)/2$.

We note that in view of (7.5.47) we obtain in both cases

(8.6) $\quad U \in S_{nd}(\Omega), \quad U \in \mathcal{C}^\infty(\Omega \setminus \Gamma)$

where Γ is given respectively by the closed nowhere dense subsets, see (6.11)

$$(8.7)\quad \begin{aligned} \Gamma_F &= \{(t,x) \in \Omega \mid x = u_l t \text{ or } x = u_r t\} \\ \Gamma_R &= \{\, (t,x) \in \Omega \mid x = st \,\} \end{aligned}$$

The First Track, Usual, Nonparametric Approach. In chapter 6, we presented the particular and simple case of defining the nonprojectable Lie group actions (8.2) on the rarefaction waves (8.4), and on the Riemann solvers (8.5), when these generalised solutions are *restricted* to convenient subsets of their natural domains of definition $\Omega = (0, \infty) \times \mathbf{R}$, see (6.32), (6.33), respectively (6.41), (6.42). These restrictions are done on the *largest* such subsets on which the respective solutions happen to be classical solutions. In this way, those results remain within the particular and classical framework of subsection 7.5.1, of Lie group actions on $\mathcal{C}^\infty$-smooth functions, rather than on generalised ones, as is the case with the constructions in subsections 7.5.2 and 7.5.3.
Therefore, the results in chapter 6 have the deficiency that they cannot deal with the rarefaction waves and the Riemann solvers *globally*, that is, on the whole of their domains of definition which are, see (8.3), given by $\Omega = (0, \infty) \times \mathbf{R}$. In particular, they do not deal with the respective *singularities* (8.7) of the rarefaction waves and Riemann solvers.
This deficiency will be remedied in the present chapter, albeit as it turns out, it cannot be done so to the full extent one may have expected, as long as one is using the usual, first track approach, see end of chapter 3, which is not the parametric representation of functions and generalised functions, that is, it is not the second track approach. Indeed, the respective limitations of the nonparametric approach prove to be deeply embedded into the situation, and as such, are rather unavoidable. An illustration of this fact was already found in Examples 4 and 5, as well as Remark 8, in subsection 7.5.3.

Here again, it is particularly important to note that these technical difficulties are *not* related to the construction of the nonprojectable Lie group symmetries of the rarefaction waves or of the Riemann solvers, construction which is rather straightforward, see $g\ \beta_*(U)$ in (8.13), (8.44) below. On the contrary, these difficulties are *solely* related to the computation of $\delta(g\ \beta_*(U))$, that is, to the computation of the whole *families* of the usual, nonparametric - and typically, only locally defined - generalised functions which can be associated with $g\ \beta_*(U)$, see comment at the beginning of subsection 7.4.7.

The second track, that is, the parametric approach in chapter 9, will therefore prove to be useful precisely to the extent that it can overcome these difficulties and limitations of the usual, nonparametric way.

Our starting point for a global treatment - including that of singularities - for the rarefaction waves and the Riemann solvers is in their expressions in (8.6), (8.7), respectively. And we shall apply to them appropriate versions of the general method in chapter 7, regarding arbitrary nonlinear, including nonprojectable, Lie group actions on generalised functions in $S_{nd}(\Omega)$, see (7.5.30) - (7.5.69).

In particular, we shall extend to two dimensions, and to the more general type of nonprojectable Lie group actions (8.2), the one dimensional cases in Examples 4 and 5, subsection 7.5.3.

Generalised Solutions Go Into Generalised Solutions. As mentioned, one of the main results in this work is the proof of the fact that the usual Lie group symmetries, including the nonprojectable ones, of classical solutions of nonlinear partial differential equations will as well remain symmetries of their generalised solutions, see Theorem 2 and Remark 11 in chapter 9, section 1. In other words, the classes upon which the usual Lie group symmetries of nonlinear equations act, that is, the classical solutions of such equations, can be *extended* to the generalised solutions of such equations, provided that these generalised solutions are in the nowhere dense differential algebras (1.6), (7.1.4), and in particular, in $S_{nd}(\Omega)$, see (7.5.46).
Since the main novelty in this work is the introduction of the *second track*, that is, *parametric* approach to Lie group actions on classical and generalised solutions, the natural basic starting result is in Theorem 1, at the end of chapter 5. There it is shown that the classical solutions given in parametric form will again turn into such solutions, when acted upon by the usual Lie group symmetries of the respective equations.

So far, however, we have not yet presented the mentioned extension of Theorem 1 to generalised functions. And this will be done in Theorem 2, in chapter 9, section 1.
Prior to doing so in general terms, in this chapter, we shall show that such an extension property clearly holds in certain particular cases of nonprojectable Lie group symmetries of the Riemann solvers. Here we can mention that this issue of extension did not arise in Examples 4, 5 in section 7.5., since the generalised functions in those examples were not taken as solutions of any partial differential equations. In particular, the nonprojectable Lie groups acting upon them were not the symmetry groups of such equations. In other words, the aim of those examples was merely to test and show how nonprojectable Lie groups act on generalised functions in $S_{nd}(\Omega)$, and to give us an insight - useful in the present chapter - about the possible difficulties involved, when one tries to keep to the first track, that is, usual, nonparametric approach.

1. Rarefaction Waves

First we start with the rarefaction waves (8.4) and represent them by, see (7.5.44)

(8.8) $$U \;=\; s + \mathcal{I}_{nd}(\Omega) \in S_{nd}(\Omega) \subseteq A_\Omega$$

where

(8.9) $$s = (\psi_\nu \mid \nu \in \mathbf{N}) \in (\mathcal{C}^\infty(\Omega))^{\mathbf{N}}$$

with, see (7.5.75)

$$(8.10) \quad \psi_\nu(t,x) = u_l + c_\nu(x/t - u_l)(x/t - u_l) - c_\nu(u_r - x/t)(x/t - u_r)$$

for $\nu \in \mathbf{N}$, $(t,x) \in \Omega = (0,\infty) \times \mathbf{R}$, thus we obtain the following relationship between ψ_ν and the rarefaction wave U in (8.4), see (7.5.77), namely, on Ω_ν we have

$$\psi_\nu = U \quad \text{on } \bar{\Omega}_{\nu 1}$$

$$(8.11) \quad u_l \leq \psi_\nu \leq u_r \quad \text{on } \bar{\Omega}_{\nu 2}$$

$$0 \leq (\psi_\nu)_x(t,x) \leq 2(K+1)/t \quad \text{on } \bar{\Omega}_{\nu 2}$$

where we denoted

$$\Omega_{\nu 1} = \Omega \setminus \left\{ (t,x) \in \Omega \;\middle|\; \begin{array}{l} u_l \leq x/t \leq u_l + 1/(\nu+1) \text{ or} \\ u_r - 1/(\nu+1) \leq x/t \leq u_r \end{array} \right\}$$

(8.12)

$$\Omega_{\nu 2} = \left\{ (t,x) \in \Omega \;\middle|\; \begin{array}{l} u_l < x/t < u_l + 1/(\nu+1) \text{ or} \\ u_r - 1/(\nu+1) < x/t < u_r \end{array} \right\}$$

Given now any $\epsilon = g \in G = (\mathbf{R}, +)$, then (7.5.52), (7.5.58) imply that

$$(8.13) \quad g\,\beta_*(U) = \text{class}_{nd}\, g s_* \in B_\Omega \subseteq B_{[\Omega]}, \quad \text{if } \Omega_{(g,\psi)} \neq \phi$$

where, see (7.5.45), (6.16)

$$(8.14) \quad \psi = U|_{\Delta_F}$$

It follows from (8.13), (7.4.46) that we can proceed and compute

$$(8.15) \quad \delta(g\,\beta_*(U))$$

as soon as we have shown that

$$(8.16) \quad \Omega_{(g,\psi)} \neq \phi$$

However, in view of (7.5.59), (8.14) , (6.17), the relation (8.16) is obviously true. Now for the computation of (8.15), let us denote for $\nu \in \mathbf{N}$, see (7.5.53)

$$(8.17) \quad V_\nu = g\psi_{\nu *} \in \mathcal{C}^\infty(\Omega, M)$$

then (7.4.19) gives the $\mathcal{C}^\infty$-smooth surjective local diffeomorphism

$$(8.18) \quad D_\nu \xrightarrow{V_{\nu 1}} \Delta'_\nu = V_{\nu 1}(D_\nu)$$

where, see (7.4.18), (6.13)

$$(8.19) \quad D_\nu = \{(t,x) \in \Omega \mid 1 + \epsilon a'(\psi_\nu(t,x))(\psi_\nu)_x(t,x) \neq 0\}$$

However, in view of (8.11), for $\nu \in \mathbf{N}$, $(t,x) \in \Omega$, we have

$$(8.20) \quad m_a/t \leq a'(\psi_\nu(t,x))(\psi_\nu)_x(t,x) \leq M_a/t$$

where $m_a, M_a \in \mathbf{R}$ do not depend on ν, t, x.

Here we shall compute certain *simplified* versions of $\delta(g\ \beta_*(U))$, see (8.15), (7.4.50), (7.4.51), (7.4.58), (7.4.63), in several cases of the nonprojectable and nonlinear Lie group symmetries (8.2), when applied to the rarefaction waves (8.4). A more complete computation in the case of the first track, usual, nonparametric approach would lead to significant technical complications, as suggested already in the simpler, one dimensional Example 4, in subsection 7.5.3.
Furthermore, such a complete computation actually becomes unnecessary, if we use the second track, parametric approach, see chapter 9.

Let us therefore, in view of (8.19), first assume the particular case, when the arbitrary function $a \in \mathcal{C}^\infty(\mathbf{R})$ in the Lie group (8.2) is *nondecreasing* on the range of the rarefaction wave, that is

$$(8.21) \quad a' \geq 0 \quad \text{on} \quad [u_l, u_r]$$

Then clearly, (8.19), (8.11) imply

$$(8.22) \quad D_\nu = \Omega \quad \text{for} \ \epsilon \geq 0 \ , \nu \in \mathbf{N}$$

Furthermore, in view of the comment following (6.13), the relations (8.18), (8.11), (8.2) give that the mapping

$$(8.23) \quad \Omega \xrightarrow{V_{\nu 1}} \Delta'_\nu = V_{\nu 1}(\Omega) = \Omega$$

is a $\mathcal{C}^\infty$-smooth diffeomorphism, for $\epsilon \geq 0$, $\nu \in \mathbf{N}$. Therefore, we are in the particular situation of pct. 4) in Remark 4, subsection 7.4.4., namely, when (7.4.59) - (7.4.63), and hence (7.4.51) apply. In this way, under a comment similar to that following (7.4.58), we can *simplify* δ in (8.15) to the form

$$(8.24) \quad \delta(g\ \beta_*(U)) = \bar{U}' \in A_\Omega, \quad \text{for} \ \epsilon = g \geq 0$$

where, see (8.17), (7.4.56), (7.4.57)

$$(8.25) \quad \bar{U}' = \text{class}_{nd}(V_{\nu 2} \circ V_{\nu 1}^{-1} \mid \nu \in \mathbf{N})$$

Let us now return to the general situation in (8.17) - (8.20) and compute δ in the two exhaustive cases, when

$$(8.26) \quad \epsilon \geq 0$$

respectively

$$(8.27) \quad \epsilon \leq 0$$

Now, if (8.26) holds, the (8.20) yields

$$1+\epsilon M_a/t \leq 1+\epsilon a'(\psi_\nu(t,x))(\psi_\nu)_x(t,x) \leq 1+\epsilon m_a/t$$

for $\nu \in \mathbf{N},\ (t,x) \in \Omega$. Hence

(8.28) $\quad (\epsilon m_a < -t \quad \text{or} \quad -t < \epsilon M_a) \Rightarrow (\ (t,x) \in D_\nu, \quad \text{for} \quad \nu \in \mathbf{N}\ , x \in \mathbf{R}\ )$

In other words

(8.29) $\quad D_\nu \supseteq I_{a\epsilon} \times \mathbf{R}, \quad \text{for} \quad \nu \in \mathbf{N}$

where we denoted the open time interval

(8.30) $\quad I_{a\epsilon} = ((0,\infty) \cap (-\infty, -\epsilon m_a)) \cup ((0,\infty) \cap (-\epsilon M_a, \infty)) \subseteq (0,\infty)$

hence clearly

(8.31) $\quad I_{a\epsilon} \neq \phi$

In view of (8.29), let us consider the open subsets in Ω given by

$$\text{(8.32)} \quad \begin{aligned} \Delta''_{a\epsilon 1} &= ((0,\infty) \cap (-\infty, -\epsilon m_a)) \times \mathbf{R} \\ \Delta''_{a\epsilon 2} &= ((0,\infty) \cap (-\epsilon M_a, \infty)) \times \mathbf{R} \end{aligned}$$

and note that obviously

(8.33) $\quad \Delta''_{a\epsilon 2} \neq \phi$

while in view of (8.2), (8.18), (8.11) and the comment after (6.13), it follows that the mapping

(8.34) $\quad \Delta''_{a\epsilon 2} \xrightarrow{V_{\nu 1}} \Delta'_\nu = V_{\nu 1}(\Delta''_{a\epsilon 2}) = \Delta''_{a\epsilon 2}$

is a $\mathcal{C}^\infty$-smooth diffeomorphism. And then, as a slight extension of the situation analyzed in Remark 4, subsection 7.4.4., we obtain that

(8.35) $\quad \delta(g\ \beta_*(U)) \supseteq (\ \bar{U}'\ )$

where the inclusion $\supseteq$ means that $\bar{U}'$ in the right hand term belongs to the family, see (7.4.49), in the left hand term, while

(8.36) $\quad \bar{U}' = \text{class}_{nd}(V_{\nu 2} \circ V_{\nu 1}^{-1} \mid \nu \in \mathbf{N}) \in A_{\Delta''_{a\epsilon 2}}$

with $V_{\nu 1}^{-1}$ computed according to (8.34).

Further, in case that

(8.37) $\quad \Delta''_{a\epsilon 1} \neq \phi$

which obviously is equivalent with

(8.38) $\quad \epsilon m_a < 0$

then similar to (8.34), we shall have that

$$(8.39) \quad \Delta''_{a\epsilon 1} \xrightarrow{V_{\nu 1}} \Delta'_{\nu} = V_{\nu 1}(\Delta''_{a\epsilon 1}) = \Delta''_{a\epsilon 1}$$

is a $\mathcal{C}^\infty$-smooth diffeomorphism, hence relations similar to (8.35), (8.36) will follow.

Finally, if (8.27) holds, then by a corresponding obvious modification of the above arguments, one can again obtain results similar to (8.35), (8.36).

In conclusion, the above results mean that, for *every* $\epsilon = g \in G = (\mathbf{R}, +)$, that is, for both cases (8.26), (8.27), one can construct the nonprojectable Lie group symmetries (8.2) of the rarefaction waves (8.4), and do so on the *whole* of their *space* domain $\mathbf{R}$, provided that, as indicated for instance in (8.32), one limits in an appropriate manner the *time* domains. This in particular implies that results such as, for instance, in (8.35), (8.36) give generalised functions whose domain contains part of the *singularities* Γ_F of the rarefaction waves (8.4).
In other words, these results, by being able to handle singularities, do indeed improve on those in chapter 6. Also, they give further particular answers to Problem 1 in subsection 7.5.2.

Let us mention here that, in the case of an arbitrary a in (8.2), a generalised solution $\bar{U}'$ in (8.23) - (8.25) need *not* again be a rarefaction wave. However, on those open subsets of its domain Ω on which it happens to be a classical function, it will by necessity be again a classical solution of the nonlinear shock wave equation (8.1), as follows from the usual Lie theory.
Furthermore, as follows from Theorem 2, in chapter 9, it will again be a generalised solution of the nonlinear shock wave equation (8.1), when considered *globally*, on the whole of Ω.

2. Riemann Solvers

We turn now to the nonprojectable Lie group symmetries (8.2) of the Riemann solvers (8.5). Let us, therefore, represent such a Riemann solver U by, see (7.5.44)

$$(8.40) \quad U \; = \; s + \mathcal{I}_{nd}(\Omega) \in S_{nd}(\Omega) \subset A_\Omega$$

where

$$(8.41) \quad s = (\psi_\nu \mid \nu \in \mathbf{N}) \in (\mathcal{C}^\infty(\Omega))^{\mathbf{N}}$$

with, see (7.5.75)

$$(8.42) \quad \psi_\nu(t, x) = u_l + (u_r - u_l)c_\nu(x - st)$$

for $\nu \in \mathbf{N}, \; (t, x) \in \Omega = (0, \infty) \times \mathbf{R}$. Hence

$$\psi_\nu(t,x) = \left| \begin{array}{ll} u_l & \text{if } x < st \\ & \\ u_r & \text{if } x > st + 1/(\nu+1) \end{array} \right.$$

(8.43)

$$u_r \leq \psi(t,x) \leq u_l$$

$$K(\nu+1)(u_r - u_l) \leq (\psi_\nu)_x(t,x) \leq 0$$

Now, similar to (8.13) - (8.16), we obtain for $\epsilon = g \in G = (\mathbf{R}, +)$ the relations

(8.44) $\quad g\ \beta_*(U) = \text{class}_{nd}\ gs_* \in B_\Omega \subseteq B_{[\Omega]}, \quad \text{if} \quad \Omega_{(g,\psi)} \neq \phi$

where, see (6.16)

(8.45) $\quad \psi = U|_{\Delta_R}, \qquad \Omega_{(g,\psi)} \neq \phi$

Therefore, we can proceed and compute, see (7.4.49)

(8.46) $\quad \delta(g\ \beta_*(U)) = (\ U_i' \mid i \in I\ )$

For that purpose, given $\epsilon = g \in G = (\mathbf{R}, +)$, let us denote for $\nu \in \mathbf{N}$

(8.47) $\quad V_\nu = g\ \psi_\nu\ {}_* \in \mathcal{C}^\infty(\Omega, M)$

then (8.2) yields for $(t,x) \in \Omega$

$$V_\nu(t,x) = (V_{\nu 1}(t,x), V_{\nu 2}(t,x)) \in M$$

(8.48) $$V_{\nu 1}(t,x) = (t, x + \epsilon\ a(\psi_\nu(t,x))) \in \Omega$$

$$V_{\nu 2}(t,x) = \psi_\nu(t,x) \in \mathbf{R}$$

and the Jacobian of $V_{\nu 1}$ is, see (8.42)

$$J_\nu(t,x) = \left| \begin{array}{ll} 1 & \epsilon a'(\psi_\nu(t,x))(\psi_\nu)_t(t,x) \\ & \\ 0 & 1 + \epsilon a'(\psi_\nu(t,x))(\psi_\nu)_x(t,x) \end{array} \right|$$

hence

$$J_\nu(t,x) = 1 + \epsilon a'(\psi_\nu(t,x))(u_r - u_l)b((\nu+1)(x - st))$$

Therefore, according to (7.4.18), we have for $\nu \in \mathbf{N}$

(8.49) $\quad D_\nu = \{(t,x) \in \Omega \mid 1 + \epsilon(u_r - u_l)a'(\psi_\nu(t,x))b((\nu+1)(x-st)) \neq 0\}$

and in view of (8.42), (8.43), it follows that

(8.50) $\quad D_\nu \supseteq \Omega \setminus \{(t,x) \in \Omega \mid 0 \leq x - st \leq 1/(\nu+1)\}$

In particular, see (6.16)

$$(8.51)\quad \begin{aligned} D_\nu \supseteq \Delta_{Rl} &= \{(t,x)\in\Omega \mid x < st\} \\ D_\nu \supseteq \Delta_{Rr\ \nu} &= \{(t,x)\in\Omega \mid x > st + 1/(\nu+1)\} \end{aligned}$$

Given now $\mu \in \mathbf{N}$, then for $\nu \in \mathbf{N}$, $\nu \geq \mu$, we obtain that, see (8.43), (8.48)

$$(8.52)\quad \begin{aligned} &\Delta_{Rl} \xrightarrow{V_{\nu 1}} \Delta'_{Rl} = V_{\nu 1}(\Delta_{Rl}) \subseteq \Omega \\ &\Delta_{Rr\ \nu} \xrightarrow{V_{\nu 1}} \Delta'_{Rr\ \nu} = V_{\nu 1}(\Delta_{Rr\ \nu}) \subseteq \Omega \end{aligned}$$

are $\mathcal{C}^\infty$-smooth diffeomorphisms, where

$$(8.53)\quad \begin{aligned} \Delta'_{Rl} &= \{(t, x+\epsilon a(u_l)) \mid x < st\} \\ \Delta'_{Rr\ \nu} &= \{(t, x+\epsilon a(u_r)) \mid x > st + 1/(\nu+1)\} \end{aligned}$$

As suggested by the situation in Example 5 and Remark 8 in subsection 7.5.3, the computation of the *whole* family of generalised functions $U'_i \in A_{\Delta'_i}$, with $i \in I$, which gives $\delta(g\ \beta_*(u))$ in (8.46), may turn out to be a rather daunting task. In view of this, here, we shall limit ourselves to pointing out some of the more relevant generalised functions in this family. For that, let us make (8.52), (8.53) more explicit, by noting that (8.48), (8.51), (8.43) imply

$$(8.54)\quad V_{\nu 1}(t,x) = \left| \begin{array}{ll} (t, x+\epsilon a(u_l)) & \text{if } x < st \\ (t, x+\epsilon a(u_r)) & \text{if } x > st + 1/(\mu+1) \end{array} \right.$$

whenever $\nu, \mu \in \mathbf{N}$, $\nu \geq \mu$, $(t,x) \in \Omega$, and $\epsilon \in \mathbf{R}$. It follows that, similar to the simpler case in (7.5.110), (7.5.111), and based on the method in subsection 7.4.8., we can use the commutative diagrams (7.4.32) and construct the generalised functions

$$(8.55)\quad \begin{aligned} U'_l &= \text{class}_{nd}(U'_{l\ \nu} = V_{\nu 2} \circ V_{\nu 1}^{-1} | \nu \in \mathbf{N}) \in A_{\Delta'_{Rl}} \\ U'_r &= \text{class}_{nd}(U'_{r\ \nu} = V_{\nu 2} \circ V_{\nu 1}^{-1} | \nu \in \mathbf{N}) \in A_{\Delta'_{Rr}} \end{aligned}$$

where, see (8.53)

$$(8.56)\quad \Delta'_{Rr} = \{(t, x+\epsilon a(u_r)) | x > st\} \subseteq \Omega$$

in which case, see (7.5.112), (8.46)

$$(8.57)\quad \delta(g\ \beta_*(U)) \supseteq (\ U'_l,\ U'_r\ )$$

and clearly, (8.54), (8.48) yield

$$(8.58)\quad \begin{aligned} U'_l &= u_l \quad \text{on } \Delta'_{Rl} \\ U'_r &= u_r \quad \text{on } \Delta'_{Rr} \end{aligned}$$

In other words, we have indentified in (8.55), (8.57) two generalised - this time, in fact, usual - functions in the family which gives $\delta(g\ \beta_*(U))$ in (8.46), and we have done so under general conditions, that is, without any restrictions related to the nonprojectable Lie group symmetries (8.2), the group parameter ϵ, or the Riemann solvers (8.5).
However, U_l', U_r' in (8.57) are *not* global, that is, they are not defined on the whole of Ω. Therefore, next, we shall try to see whether in the family which gives $\delta(g\ \beta_*(U))$, we may find usual, *nonparametric* generalised functions defined *globally*, on the whole of Ω.

Related to these two functions U_l', U_r' in (8.55), we have the following three cases with respect to the mutual position of their domains of definition, see also (6.16), (6.35) - (6.43), (8.53)

(8.59) $\quad \Delta'_{Rl} \cap \Delta'_{Rr} = \phi \quad \text{and} \quad \epsilon a(u_l) < \epsilon a(u_r)$

(8.60) $\quad \Delta'_{Rl} \cap \Delta'_{Rr} = \phi \quad \text{and} \quad \epsilon a(u_l) = \epsilon a(u_r) = x_\epsilon$

(8.61) $\quad \Delta'_{Rl} \cap \Delta'_{Rr} \neq \phi \quad \text{and} \quad \epsilon a(u_l) > \epsilon a(u_r)$

It is important to note regarding theses three possible cases (8.59) - (8.61) that, given the nonprojectable Lie group symmetry (8.2) of the nonlinear shock wave equation (8.1), that is, given the $\mathcal{C}^\infty$-smooth function a, the three cases above only depend on u_l, u_r and ϵ. In particular, if the Riemann solver U in (8.5), and thus u_l and u_r are also given, then it will be the Lie group parameter ϵ alone which will determine which of the three cases (8.59) - (8.61) may apply.

And now, let us study each of these three cases.
Case (8.60) is easy to deal with, since it implies that the subset

(8.62) $\quad \Gamma_{R\,\epsilon} \;=\; \Omega \setminus (\Delta'_{Rl} \cup \Delta'_{Rr}) \;=\; \{(t,x) \in \Omega \mid x = st + x_\epsilon\}$

is closed and nowhere dense in Ω. Therefore, in view of (8.58), (7.5.47), there exists a unique generalized function $U' \in \mathcal{S}_{nd}(\Omega)$, such that U_l', U_r' are restrictions of U' to their respective domains of definition Δ'_{Rl} and Δ'_{Rr}. Moreover, in view of (8.62), we have

(8.63) $\quad U'(t,x) \;=\; u_l + (u_r - u_l)H(x - st - x_\epsilon), \quad (t,x) \in \Omega$

It follows that the generalised function $U' \in \mathcal{S}_{nd}(\Omega)$ in (8.63) is *itself* a Riemann solver for the nonlinear shock wave equation (8.1), which this time corresponds to the *shifted* initial condition, see (6.2)

(8.64) $\quad U'(0,x) = u_l + (u_r - u_l)H(x - x_\epsilon), \quad x \in \mathbf{R}$

Here it is important to note that, since $u_l > u_r$, the Riemann solver (8.63) is the *only global* generalised solution of the nonlinear shock wave equation (8.1), which is defined on the whole of Ω, and such that, see Smoller

$$U' = U'_l \quad \text{on} \quad \Delta'_{Rl}$$

(8.65) $$U' = U'_r \quad \text{on} \quad \Delta'_{Rr}$$

U' satisfies the Rankine-Hugoniot and the entropy conditions

In short, U' in (8.63) is the *only* physically meaningful solution corresponding to the initial condition (8.64).
At the end of this chapter, we shall show that in the case (8.60), we can indeed improve on (8.57) by the *global* result

(8.66) $$\delta(g\ \beta_*(U)) \supseteq (\ U'\ )$$

Let us turn now to the case (8.59). We consider the nonvoid open subset of Ω, given by

(8.67) $$\Delta'_{Rc} = \{\ (t,x) \in \Omega \mid st + \epsilon a(u_l) < x < st + \epsilon a(u_r)\ \}$$

then clearly, the nonvoid open subsets of Ω

(8.68) $$\Delta'_{Rl},\ \Delta'_{Rc},\ \Delta'_{Rl} \quad \text{are pairwise disjoint}$$

and

(8.69) $$\Gamma_{Rl\ \epsilon} \cup \Gamma_{Rr\ \epsilon} = \Omega \setminus (\Delta'_{Rl} \cup \Delta'_{Rc} \cup \Delta'_{Rr})$$

where

$$\Gamma_{Rl\ \epsilon} = \{\ (t,x) \mid x = st + \epsilon a(u_l)\ \}$$

(8.70)

$$\Gamma_{Rr\ \epsilon} = \{\ (t,x) \mid x = st + \epsilon a(u_r)\ \}$$

are closed and nowhere dense subsets in Ω.

Unlike in case (8.60), this time the problem is in the *gap* created by the nonvoid open subset Δ'_{Rc} between the respective domains Δ'_{Rl} and Δ'_{Rr} of the generalised functions U'_l, U'_r in (8.55).
However, it is easy to see that the two closed nowhere dense subsets $\Gamma_{Rl\ \epsilon}$, $\Gamma_{Rr\ \epsilon}$ in (8.70) are two parallel half lines. In view of that, we shall be able to bridge the mentioned gap in an easy manner. Namely, let us take any x_0 such that

(8.71) $$\epsilon a(u_l) \leq x_0 \leq \epsilon a(u_r)$$

and define the Riemann solver

(8.72) $$U'_0(t,x) = u_l + (u_r - u_l)H(x - st - x_0), \quad (t,x) \in \Omega$$

which therefore corresponds to the following shifted initial value for the nonlinear shock wave equation (8.1)

(8.73) $$U'_0(0,x) = u_l + (u_r - u_l)H(x - x_0), \quad x \in \mathbf{R}$$

Then the shock front, that is, the singularity set of U'_0 is given by the closed nowhere dense subset

$$(8.74) \quad \Gamma_{R\, x_0} = \{(t,x) \in \Omega \mid x = st + x_0\}$$

and for each x_0 in (8.71), the corresponding Riemann solver $U'_0 \in S_{nd}(\Omega)$ given by (8.72), will satisfy the conditions (8.65). It follows that, again, and just as in (8.63), the Riemann solver U'_0 in (8.72) is the *only* physically meaningful solution which, this time, corresponds to the initial condition (8.73).
Furthermore, the family of Riemann solvers (8.72), (8.71) are the *only global* generalised solutions of the nonlinear shock wave equation (8.1) which are defined on the whole of Ω and also satisfy (8.65).

However, in spite of (8.71) - (8.74), it appears that in the case (8.59) we *cannot* improve on (8.57) by a *global* result of the form

$$(8.75) \quad \delta(\epsilon\, \beta_*(U)) \supseteq (\, U'_0 \mid x_0 \in [\epsilon a(u_l), \epsilon a(u_r)]\,)$$

unless additional special conditions are satisfied by the function a in (8.2), and thus in general, we are only left with the result in (8.57).

Finally, let us turn to case (8.61).
Similar with (8.67), let us now denote

$$(8.76) \quad \Delta'_{Rc} = \Delta'_{Rl} \cap \Delta'_{Rr} = \{\, (t,x) \in \Omega \mid st + \epsilon a(u_r) < x < st + \epsilon a(u_l)\, \}$$

This time the nonvoid open subset Δ'_{Rc} is not causing a gap between the domains of definition of the two generalised functions in (8.55), but on the contrary, it describes their *overlap*.
However, let us recall that $u_l > u_r$, and also (8.58), as well as the fact that the boundaries between $\Delta'_{Rl} \setminus \Delta'_{Rc}$ and Δ'_{Rc}, respectively, between $\Delta'_{Rr} \setminus \Delta'_{Rc}$ and Δ'_{Rc}, are given by the half lines $\Gamma_{Rr\, \epsilon}$ and $\Gamma_{Rl\, \epsilon}$ in (8.70).
It follows that, we can take any, see (8.71)

$$(8.77) \quad \epsilon a(u_r) \leq x_0 \leq \epsilon a(u_l)$$

and define the corresponding version of the Riemann solver in (8.72), see also (8.73), (8.74)

$$(8.78) \quad U'_0(t,x) = u_l + (u_r - u_l) H(x - st - x_0), \quad (t,x) \in \Omega$$

And again, the family (8.78) of Riemann solvers are the *only global* generalised solutions of the nonlinear shock wave equation (8.1), which are defined on the whole of Ω, and which satisfy, see (8.65)

$$(8.79) \quad \begin{aligned} & U'_0 = U'_l \quad \text{on } \Delta'_{Rl} \setminus \Delta'_{Rc} \\ & U'_0 = U'_r \quad \text{on } \Delta'_{Rr} \setminus \Delta'_{Rc} \\ & U'_0 \quad \text{satisfies the Rankine-Hugoniot and entropy conditions} \end{aligned}$$

In particular, U_0' is once again the *only* physically meaningful solution for the initial conditions (8.73). And its *singularity* set is given in (8.74).
Again, as in (8.59), (8.75), it appears that in the case (8.61), we cannot in general improve on (8.57) by the *global* result, see (8.75)

$$(8.80) \quad \delta(g\ \beta_*(U)) \supseteq (\ U_0' \mid x_0 \in [\epsilon a(u_r), \epsilon a(u_l)]\ )$$

We can now recapitulate.
Given the Riemann solver (8.4), with $u_l > u_r$, let us apply to it the nonprojectable Lie symmetry group (8.2) of the nonlinear shock wave equation (8.1). Then we have the three possibilities

$$(8.81) \quad a(u_l) < a(u_r)$$

$$(8.82) \quad a(u_l) = a(u_r)$$

$$(8.83) \quad a(u_l) > a(u_r)$$

Let us take now any value $\epsilon \in \mathbf{R}$, $\epsilon \neq 0$, of the Lie group parameter.
First, let us assume that $\epsilon > 0$. Then (8.81), (8.82) and (8.83) will respctively lead to the cases (8.59), (8.60) and (8.61).
Conversely, if $\epsilon < 0$, then (8.81), (8.82) and (8.83) will respectively imply the cases (8.61), (8.60) and (8.59).

Let us now focus, as above, on the physically meaningful global generalised solutions of the nonlinear shock wave equation (8.1), that is, on those solutions which are defined on the whole of Ω, and also satisfy the conservation type Rankine-Hugoniot, and the entropy conditions, see (8.65), (8.79).

In the case (8.60), we shall then obtain the *unique* Riemann solver U' in (8.66).
The other two cases (8.59) and (8.61) will lead us to the respective *families* of Riemann solvers in (8.75) and (8.80).
However, as far as the values of δ are concerned, it appears that we shall in general only have (8.66), which corresponds to the case (8.60), and when these values contain as well a *global* Riemann solver. Otherwise, we have the general situation in (8.57).

Remark 10

1) If, see (8.82)

$$(8.84) \quad a(u_l) = a(u_r)$$

then for all $\epsilon \in \mathbf{R}$, the result of the nonprojectable Lie group action (8.2) on the Riemann solver (8.5) is the *unique* Riemann solver in (8.66).

2) A certain *closure* property of the mapping δ can be observed in the results in this section.
Indeed, in view of (8.45), for every $\epsilon \in \mathbf{R}$, the respective nonprojectable Lie group action (8.2) can be performed on every Riemann solver (8.5) of the nonlinear shock wave equation (8.1).

On the other hand, in the case (8.84), and in view of (8.66), the result of such actions are again Rieman solvers. Therefore, in such situations the iterations requested at pct. 2) in Problem 2, in section 7.7, can be performed without any restrictions.

Let us take any $\epsilon' \in \mathbf{R}$ and apply the corresponding nonprojectable Lie group action (8.2) to the Riemann solvers in (8.66). In this respect, we note that, in view of (8.58), the case in (8.82) will remain unchanged.
Let us consider the corresponding unique Riemann solver U' in (8.66). Then repeating, this time for ϵ', the earlier argument which, for ϵ, has lead us to (8.66), we shall now obtain

$$(8.85) \quad \delta(\epsilon' \ \beta_*(U')) \supseteq (\ U''\ )$$

where, see (8.63)

$$(8.86) \quad U''(t,x) \ = \ u_l + (u_r - u_l)H((x - x_\epsilon) - x_{\epsilon'}), \quad (t,x) \in \Omega$$

with, see (8.60)

$$(8.87) \quad x_{\epsilon'} = \epsilon' \ a(u_l) = \epsilon' \ a(u_r)$$

However, according to the group composition law in (8.2), we have

$$(8.88) \quad \epsilon'' = \epsilon' \ \epsilon = \epsilon + \epsilon'$$

therefore (8.60), (8.87) and (8.88) give

$$(8.89) \quad x_{\epsilon''} = x_\epsilon + x_{\epsilon'}$$

and then (8.86) becomes

$$(8.90) \quad U''(t,x) \ = \ u_l + (u_r - u_l)H(x - x_{\epsilon''}), \quad (t,x) \in \Omega$$

In this way, in the terms of (7.7.2), (7.7.3), we do have the *closure* property

$$\begin{aligned} &(\ U'\ ) \subseteq \delta(\epsilon \ \beta_*(U)) \\ (8.91) \qquad &(\ U''\ ) \subseteq \delta(\epsilon' \ \beta_*(U')) \\ &(\ U''\ ) \subseteq \delta(\epsilon'' \ \beta_*(U)) \end{aligned}$$

□

And now, let us return to the proof of (8.66), which corresponds to the case (8.60). In view of (8.52), it is obvious that the mappings

$$(8.92) \quad \Delta_{Rl} \cup \Delta_{Rr\nu} \xrightarrow{V_\nu} \Delta'_{Rl} \cup \Delta'_{Rr\nu}, \quad \nu \in \mathbf{N}$$

are $\mathcal{C}^\infty$-smooth diffeomorphisms, thus by the use of the commutative diagrams (7.4.32), we can obtaon the sequence of functions

(8.93) $\quad U'_\nu = V_{\nu 2} \circ V'_{\nu 1} \in \mathcal{C}^\infty(\Delta'_{Rl} \cup \Delta^{p}rime_{Rr\nu}), \quad \nu \in \mathbf{N}$

and by applying the method in (7.4.111) - (7.4.119), we obtain (8.66).

We can obtain (8.66) in the following alternative way as well. Let us replace the sequence of functions V_ν, with $\nu \in \mathbf{N}$, in (8.47), with the sequence of functions $\bar{V}_\nu$, for $\nu \in \mathbf{N}$, where

$$\text{(8.94)} \qquad \begin{aligned} &\bar{V}_{\nu 1}(t,x) = (t, x + x_\epsilon), \quad (t,x) \in \Omega \\ &\bar{V}_{\nu 2} = V_{\nu 2} \end{aligned}$$

Then clearly

(8.95) $\quad g\ \beta_*(U) = \text{class}_{nd}(\bar{V}_\nu | \nu \in \mathbf{N})$

and

(8.96) $\quad \bar{V}_\nu : \Omega \longrightarrow \Omega, \quad \nu \in \mathbf{N}$

are $\mathcal{C}^\infty$-smooth diffeomorphisms. Now if we apply to them the commutative diagrams (7.4.32), we obtain U' in (8.63), and therefore (8.66) as well.

Chapter 9

General Parametric Approach to Lie Symmetries

As mentioned - and also motivated - in section 7.6, as well as in other earlier places, the main aim of this chapter is to give an extension of the result in Theorem 1, chapter 5, on the usual Lie group symmetries of parametric classical solutions, to *parametric generalised solutions.*

And as seen in (7.3.14), (7.6.15), (7.6.16), the most general setup which appears naturally in this regard is of Lie group actions, see comment at the end of section 7.6. for the definition of the quotient space below

$$(9.1) \quad G \times ((\mathcal{C}_n^\infty(M))^{\mathbf{N}} / \approx_{nd}) \xrightarrow{\mathcal{L}_{prm}} (\mathcal{C}_n^\infty(M))^{\mathbf{N}} / \approx_{nd}$$

which in view of (4.1) - (4.4) can easily be defined precisely in the same way as their more particular forms in (7.3.14) were defined.

It follows that this chapter is dedicated to the *second track* approach, see the end of chapter 3, which will be pursued here in its full generality.

An effect of the developments in this chapter will in particular be to include - and also clarify and simplify - the results in chapters 6 - 8 on arbitrary nonlinear, including nonprojectable Lie group actions on usual, nonparametric generalised functions. This will also include the earlier results on the nonprojectable Lie group symmetries of the rarefaction waves and Riemann solvers of the nonlinear shock wave equation.

On the general level, the results in this chapter extend those in Oberguggenberger & Rosinger [pp. 315-325], where within the first track, usual nonparametric approach, and only for projectable Lie group actions, it was shown that the usual Lie group symmetries of nonlinear PDEs when applied to generalised solutions give again generalised solutions. A short account of these results is presented in chapter 10.

Let us now turn to the $\mathcal{C}^\infty$-smooth nonlinear partial differential equations in (1.1), (1.4), and to their usual Lie group symmetries in (1.2), (1.3).

1. Symmetries of Parametric Generalised Solutions

Parametric Generalised Solutions. Before going further, we have to define what we mean by a *parametric generalised solution* of (1.1), (1.4). And naturally, this definition has to contain as a particular case the respective definition (5.42) of a parametric classical solution.
Given therefore

(9.1.1) $\quad V \in (\mathcal{C}_n^\infty(M))^{\mathbf{N}} / \approx_{nd}$

let us assume that

(9.1.2) $\quad V = \text{class}_{nd}\,(V_\nu | \nu \in \mathbf{N})$

where

(9.1.3) $\quad V_\nu \in \mathcal{C}^\infty(\Lambda, M), \quad \text{for } \nu \in \mathbf{N}$

for a suitable nonvoid open subset $\Lambda \subseteq \mathbf{R}^n$.
Given now any $\nu \in \mathbf{N}$, we can clearly use (5.40), (5.41) and compute for V_ν

(9.1.4) $\quad S(y, D)\, V_\nu(y) = F(\, V_{\nu 1}(y),\ V_{\nu 2}(y),\ \ldots\,), \quad y \in D_{V_\nu} \subseteq \Lambda \subseteq \mathbf{R}^n$

Here however, we face *two* problems.

First we note that, this time, and unlike in (5.42), we cannot expect that the expression in (9.1.4) may vanish for any given $\nu \in \mathbf{N}$, since V_ν is but only one term in just one of the many possible representations (9.1.2) of the parametric generalised function V in (9.1.1).

The second problem is that the functions $S(y, D)\ V_\nu$, with $\nu \in \mathbf{N}$, in (9.1.4) have domains of definition $D_{V_\nu} \subseteq \Lambda$, which need not be identical.

However, since we are talking about generalised, and not classical solutions, what we are expecting is that, in case $D_{V_\nu} \neq \phi$, for $\nu \in \mathbf{N}$, then the sequence of parametric functions

(9.1.5) $\quad (\, S(y, D)\ V_\nu \mid \nu \in \mathbf{N}\,) \in (\mathcal{C}_n^\infty(M))^{\mathbf{N}}$

generated by (9.1.4), may vanish in a certain generalised sense, that is

(9.1.6) $\quad (\, S(y, D)\ V_\nu \mid \nu \in \mathbf{N}\,) \approx 0$

Here, for the sake of simplicity, we shall limit ourself to the precise definition of this relation $\approx$ in the special contex of (9.1.6) only. In section 4 of this chapter, we shall present a fully detailed general version of it. The difficulty in doing so is a consequence of the second problem mentioned above, namely, that the domains of definition of the functions in the sequence (9.1.5) may fail to be identical.

And now, let us return to the issue of an appropriate definition of (9.1.6). A natural way to proceed is as follows. Given $(V_\nu | \nu \in \mathbf{N})$ as in (9.1.2), let us define the possibly void open set, see (7.4.18)

$$(9.1.7) \quad D_{(V_\nu|\nu\in\mathbf{N})} = \left\{ y \in \Lambda \;\middle|\; \begin{array}{l} \exists\ \Lambda' \subseteq \Lambda \text{ neighbourhood of } y,\ \mu \in \mathbf{N} : \\ \forall\ \nu \in \mathbf{N},\ \nu \geq \mu : \\ \quad \Lambda' \subseteq D_{V_\nu} \end{array} \right\}$$

Clearly, in the case of a parametric classical solution V, when we can assume in (9.1.2) that

$$(9.1.8) \quad V = V_0 = V_1 = V_2 = \ \ldots\ = V_\nu = \ \ldots\ , \quad \nu \in \mathbf{N}$$

then (9.1.7) reduces to the earlier definition of D_V in (5.33).

Returning to the general situation in (9.1.1) - (9.1.5), we can define (9.1.6) as being equivalent with the condition

$$(9.1.9) \quad \begin{array}{l} \exists\ \Gamma \subset \Lambda, \text{ closed, nowhere dense } : \\ \forall\ y \in D_{(V_\nu|\nu\in\mathbf{N})} \setminus \Gamma : \\ \exists\ \Lambda' \subseteq \Lambda \setminus \Gamma \text{ neighbourhood of } y,\ \mu \in \mathbf{N} : \\ \forall\ \nu \in \mathbf{N},\ \nu \geq \mu : \\ \quad *)\ \Lambda' \subseteq D_{V_\nu} \\ \quad **)\ S(z, D)\ V_\nu(z) = 0, \quad z \in \Lambda' \end{array}$$

We note here that, similar with the situation of the equation (7.5.19), when seen through the associated commutative diagram (7.5.23), here also, owing to the arbitrary closed and nowhere dense subset $\Gamma \subset \Lambda$ which appears in (9.1.9), and which may depend on the sequence $(V_\nu | \nu \in \mathbf{N})$, it follows that in (9.1.6) we cannot a priori specify any particular subset of Λ, on which this relation can be stated to hold pointwise. However, as before with the nowhere dense differential algebras (1.6), see also (7.5.23), we can say that (9.1.6) holds on the open subset $D_{(V_\nu|\nu\in\mathbf{N})} \subseteq \Lambda$, modulo closed, nowhere dense subsets. This issue will be clarified in section 4 of this chapter, once the general setup which comprises the relation $\approx$ in (9.1.6) is presented there.
However, this situation in (9.1.6), which is unlike in (5.42) where the equality does hold pointwise on D_V, is in fact quite natural, since here we are no longer talking about parametric classical solutions, but about parametric *generalised* solutions.

On the other hand, it is easy to see that in the case of a parametric classical solution, that is, when (9.1.8) holds, the above condition (9.1.9) reduces to (5.42).

Fortunately however, in the general situation of (9.1.1) - (9.1.5), owing to the properties of the equivalence relation $\approx_{nd}$ in (7.3.3), as extended at the end of section 7.6., we can obtain a significant independence of $D_{(V_\nu|\nu\in\mathbf{N})}$ in (9.1.7), of the particular representation of V chosen in (9.1.2). Indeed, let us assume that we are given another such representation

$$(9.1.10) \quad V = \text{class}_{nd}\,(\bar{V}_\nu|\nu\in\mathbf{N}), \quad \text{where} \quad \bar{V}_\nu \in \mathcal{C}_n^\infty(M), \quad \text{for} \quad \nu\in\mathbf{N}$$

Then we obtain a closed nowhere dense subset $\Gamma\subset\Lambda$, such that

$$(9.1.11) \quad \begin{array}{l} \forall\, y\in\Lambda\setminus\Gamma \;: \\ \exists\, \Lambda'\subseteq\Lambda\setminus\Gamma \text{ neighbourhood of } y,\ \mu\in\mathbf{N} \;: \\ \forall\, \nu\in\mathbf{N},\ \nu\geq\mu \;: \\ \quad \bar{V}_\nu = V_\nu \ \text{ on } \Lambda' \end{array}$$

Now let us take any $y\in D_{(V_\nu|\nu\in\mathbf{N})}$. In case we have $y\in\Lambda\setminus\Gamma$, then clearly, (9.1.7), (9.1.11) will give a neighbourhood $\Lambda'\subseteq\Lambda\setminus\Gamma$ of y, and also $\mu\in\mathbf{N}$, such that

$$y\in\Lambda'\setminus\Gamma\subseteq D_{V_\nu}\cap D_{\bar{V}_\nu}, \quad \nu\in\mathbf{N}, \quad \nu\geq\mu$$

therefore $y\in D_{(\bar{V}_\nu|\nu\in\mathbf{N})}$. In this way it follows that

$$(9.1.12) \quad D_{(V_\nu|\nu\in\mathbf{N})}\setminus\Gamma = D_{(\bar{V}_\nu|\nu\in\mathbf{N})}\setminus\Gamma$$

and in particular, since Γ is closed and nowhere dense, we obtain that

$$(9.1.13) \quad D_{(V_\nu|\nu\in\mathbf{N})}\neq\phi \iff D_{(\bar{V}_\nu|\nu\in\mathbf{N})}\neq\phi$$

An important consequence of the above is the following. Let us take, see (7.5.44) - (7.5.48), any usual, nonparametric generalised function $U\in S_{nd}(\Omega)$, and associate with it, see (7.5.50), (7.3.6), (7.3.16)

$$(9.1.14) \quad \beta_*(U) = \text{class}_{nd}s_* \in (\mathcal{C}_n^\infty(M))^{\mathbf{N}}/\approx_{nd}$$

then it follows easily that, see also (9.1.40) below for the general situation

$$(9.1.15) \quad D_{s_*} = \Omega$$

therefore, according to (9.1.12), for every parametric representation $(V_\nu|\nu\in\mathbf{N})\in(\mathcal{C}_n^\infty(M))^{\mathbf{N}}$ of $\beta_*(U)$, that is

$$(9.1.16) \quad \beta_*(U) = \text{class}_{nd}(V_\nu|\nu\in\mathbf{N})$$

we shall have

$$(9.1.17) \quad D_{(V_\nu|\nu\in\mathbf{N})}\supseteq\Omega\setminus\Gamma\neq\phi$$

for a suitable closed, nowhere dense subset $\Gamma\subset\Omega$.

Now we can return to our aim here, which is to define when the parametric generalised function V in (9.1.1) is a solution of the nonlinear PDE in (1.1), (1.4).

Let us note in this regard that, even in (5.42), where the definition of a parametric classical solution V is given, such as for instance we have in (9.1.8), one cannot expect that the solution will be valid on the whole domain of definition Λ of V, but only on the open subset $D_V \subseteq \Lambda$. Furthermore, as mentioned after (9.1.9), in (9.1.6) we face a further natural loss of domain specification, as we can no longer specify any domain in Λ on which (9.1.6) may hold pointwise, except if it is only assumed to hold modulo closed, nowhere dense subsets.

Let us therefore give the following main definition.
We call V in (9.1.1) a *parametric generalised solution* of the nonlinear PDE in (1.1), (1.4), if and only if

$$(9.1.18) \quad \begin{array}{l} \exists \ (V_\nu | \nu \in \mathbf{N}) \text{ in } (9.3) \ : \\ \quad (S(y, D) \ V_\nu \mid \nu \in \mathbf{N}) \ \approx \ 0 \end{array}$$

Clearly, this definition contains as a particular case the one in (5.42).
Moreover, if V in (9.1.1) is a parametric generalised solution of the nonlinear PDE in (1.1), (1.4), then it has the property that

$$(9.1.19) \quad \begin{array}{l} \forall \ (V_\nu | \nu \in \mathbf{N}) \text{ in } (9.3) \ : \\ \quad (S(y, D) \ V_\nu \mid \nu \in \mathbf{N}) \ \approx \ 0 \end{array}$$

in other words, the fact that V is a parametric generalised solution of (1.1), (1.4), does *not* depend on the particular representation of V in (9.1.2), although it is sufficient to be established only for one such representation.
Indeed, (9.1.18), (9.1.6), (9.1.9) give a closed, nowhere dense $\Gamma \subset \Lambda$, such that

$$(9.1.20) \quad \begin{array}{l} \forall \ y \in D_{(V_\nu | \nu \in \mathbf{N})} \setminus \Gamma \ : \\ \exists \ \Lambda' \subseteq \Lambda \setminus \Gamma \text{ neighbourhood of } y, \ \mu \in \mathbf{N} \ : \\ \forall \ \nu \in \mathbf{N}, \ \nu \geq \mu \ : \\ \quad *) \ \Lambda' \subseteq D_{V_\nu} \\ \quad **) \ S(z, D) \ V_\nu(z) = 0, \quad z \in \Lambda' \end{array}$$

Let us take now another representation (9.1.2) of V in (9.1.1), such as for instance in (9.1.10). Then (9.1.11), (9.1.12), (9.1.20) will lead to (9.1.19).

In view of (9.1.19), let us denote for V in (9.1.1)

$$(9.1.21) \quad D_V \ = \ \bigcup D_{(V_\nu | \nu \in \mathbf{N})}$$

where the union is taken over all representations (9.1.2) of V. In view of (9.1.13), obviously

$$(9.1.22) \quad D_V \neq \phi \Longleftrightarrow D_{(V_\nu | \nu \in \mathbf{N})} \neq \phi$$

where $(V_\nu | \nu \in \mathbf{N})$ is just one given representation (9.1.2) of V in (9.1.1).

In particular, for any usual, nonparametric generalised function $U \in S_{nd}(\Omega)$, we obtain from (9.1.15) the relation, see (7.5.44) - (7.5.48)

(9.1.23) $\quad D_{\beta_\bullet(U)} = \Omega$

Returning to the general situation in (9.1.1) - (9.1.3), we note that for a given parametric generalised function V, the open set D_V in (9.1.21) does *not* depend on any particular representation (9.1.2) of V. Taking this into account, as well as property (9.1.22) of D_V, we can now reformulate the *terminology* used in definition (9.1.18), by calling the respective V a *parametric generalised solution* of (1.1), (1.4), *on the open set* D_V.

Note on the Concept of Solution. It is important to note here the following. The main advantage of the parametric approach to arbitrary nonlinear, including nonprojectable Lie group actions on generalised functions is that these actions can be defined through (9.1), and this definition is in fact an easy, completely unrestricted, *global* and explicit construction, due to the functorial advantage enjoyed by (4.1) - (4.4).
On the other hand, when we come to the concept of parametric solution of a nonlinear PDE, be it a classical solution as in (5.42), or a generalised one as in (9.1.18), certain restrictions appear, owing to the fact that the open domains on which these solutions may hold can happen to be void.
However, as we have already noted, such restrictions are natural, since any nontrivial concept of solution must be rather restrictive. Moreover, such restrictions do not at all impact on the freedom and ease of the global and explicit actions in (9.1). Finally, as we have seen in chapter 5 and in (9.1.23), the concept of parametric generalised solution in (9.1.18) does contain as a particular case both the usual, nonparametric, as well as the parametric concepts of classical solutions.

Invariance of Parametric Generalised Solutions. Now we can come to the main aim of this chapter, and show that a parametric generalised solution V of the nonlinear PDE in (1.1), (1.4), will remain a parametric generalised solution of that equation, whenever it is acted upon by the usual Lie group symmetries of the mentioned equation.
Let us therefore assume given such a parametric generalised solution V in (9.1.1) as well as one of its representations $(V_\nu | \nu \in \mathbf{N})$ in (9.1.2), for which then (9.1.18) will hold.
Further, let G in (1.2) be the usual Lie symmetry group of the nonlinear PDE in (1.1), (1.4).
Given now any $g \in G$, we can define its action $g\ V$ on V by, see (4.1) - (4.4), (7.3.9) - (7.3.11), (7.3.14), (7.6.16), (9.1)

(9.1.24) $\quad g\ V \ = \ \text{class}_{nd}(gV_\nu \mid \nu \in \mathbf{N})$

where

(9.1.25) $\quad gV_\nu = g \circ V_\nu : \Lambda \to M$

All that is left now is to show that $g\ V$ in (9.1.24) does also satisfy (9.1.18). For that, let us use $(gV_\nu | \nu \in \mathbf{N})$ in (9.1.24) as a representation, see (9.1.1), (9.1.2), of $g\ V$. But $(V\nu | \nu \in \mathbf{N})$ satisfies (9.1.18), since we assumed that it is a representation (9.1.2) of V in (9.1.1). Therefore, applying (9.1.9) to it, we obtain for a suitable closed, nowhere dense subset Γ of Λ that

$$\forall\ y \in D_{(V_\nu | \nu \in \mathbf{N})} \setminus \Gamma \ :$$

$$\exists\ \Lambda' \subseteq \Lambda \setminus \Gamma \text{ neighbourhood of } y,\ \ \mu \in \mathbf{N} \ :$$

(9.1.26) $$\forall\ \nu \in \mathbf{N},\ \ \nu \geq \mu \ :$$

$$*)\ \Lambda' \subseteq D_{V_\nu}$$

$$**)\ S(z, D)\ V_\nu(z) = 0, \quad z \in \Lambda'$$

Now we take the above relation $**)$, that is

$$S(z, D)\ V_\nu(z) = 0, \quad z \in \Lambda'$$

apply to it Theorem 1, in chapter 5, and obtain, see (5.73)

(9.1.27) $\quad S(z, D)(gV_\nu)(z) = 0, \quad z \in \Lambda' \cap D_{gV_\nu}$

provided that

(9.1.28) $\quad \Lambda' \cap D_{gV_\nu} \neq \phi$

In view of $*)$ in (9.1.26), a necessary condition for (9.1.28) is that

(9.1.29) $\quad D_{V_\nu} \cap D_{gV_\nu} \neq \phi$

However, in its appropriate version presented next in (9.1.30), this also proves to be a sufficient condition for (9.1.28), more precisely, for (9.1.27).

Indeed, let us assume that, see (5.74), (9.1.18), (9.1.9), (9.1.6)

(9.1.30) $\quad D_{(V_\nu | \nu \in \mathbf{N})} \cap D_{(gV_\nu | \nu \in \mathbf{N})} \neq \phi$

then (9.1.7) implies that

$$\exists\ \Lambda'_0 \subseteq \Lambda \text{ nonvoid open},\ \ \mu \in \mathbf{N} \ :$$

(9.1.31) $$\forall\ \nu \in \mathbf{N},\ \ \nu \geq \mu \ :$$

$$\Lambda'_0 \subseteq D_{V_\nu} \cap D_{gV_\nu}$$

Therefore, in (9.1.26), we can take

$$y \in (D_{(V_\nu | \nu \in \mathbf{N})} \cap D_{(gV_\nu | \nu \in \mathbf{N})}, \quad \Lambda' = \Lambda'_0 \setminus \Gamma$$

in which case, in view of (9.1.31), it is clear that (9.1.27) will also hold.

Finally we can note that in view of (9.1.12), (9.1.13), (9.1.21), (9.1.24), the condition (9.1.30) can be written in the equivalent form

(9.1.32) $\quad D_V \cap D_{gV} \neq \phi$

which has the advantage that it is solely the terms of V and $g\ V$, and thus not in terms of any of their particular representations.

In this way, as the basic result of the parametric approach to Lie group actions on generalised functions, when in particular, they act as symmetries of generalised solutions of nonlinear PDEs, we can state the following extension of Theorem 1, in chapter 5.

Theorem 2

Let $V \in (\mathcal{C}_n^\infty(M))^{\mathbf{N}} / \approx_{nd}$ be a parametric generalised solution of the nonlinear PDE in (1.1), (1.4), on the open set D_V, and let G in (1.2) be the usual Lie group of symmetry of that equation.
Given any $g \in G$, if

(9.1.33) $\quad D_V \cap D_{gV} \neq \phi$

then the parametric generalised function $g\ V \in (\mathcal{C}_n^\infty(M))^{\mathbf{N}} / \approx_{nd}$, which is the result of the action of g on V, will again be a parametric generalised solution of the PDE in (1.1), (1.4), this time, on the open set $D_V \cap D_{gV}$.

□

Related to the condition (9.1.33), or its equivalent (9.1.30), we can recall the comments following Theorem 1 in chapter 5.

Remark 11

Theorem 2 above can be reformulated as follows, in less technical terms :

The usual Lie group symmetries of classical solutions of nonlinear PDEs are also symmetries of their parametric generalised solutions.

Four Concepts of Solution. For clarification, as well as further use in section 2, we recall here that four concepts of solution for the general *nonlinear* PDEs in (1.1), (1.4), are employed in this work : two concepts for each of the first, and the second track approaches, respectively, see diagram below.

As far as the first track approach is concerned, we deal with usual, nonparametric solutions, be it in their classical $\mathcal{C}^\infty$-smooth version, or on the contrary, in their generalised one, as given for instance by generalised functions in the nowhere dense differential algebras (1.6).

In the *linear* case, not specifically dealt with in this work, the second of these two concepts of solution, that is, the usual, nonparametric concept of generalised solution, has a long established history, starting back in the 1930s, with the work of Sobolev, and brought into its full generality by Schwartz, about a decade later.

In the *nonlinear* case, the concept of usual, nonparametric generalised solution - within nonlinear algebraic theories of generalised functions - has been developed since the 1960s, see Rosinger [1-12], Rosinger & Rudolph, Rosinger & Walus [1,2], Colombeau, Oberguggenberger, and the literature cited there. Within theories of generalised functions based on the order completion of spaces of smooth functions, a certain version of solution for arbitrary nonlinear continuous PDEs, replacing the usual, nonparametric concept of generalised solution, has been developed since the early 1990s, see Oberguggenberger & Rosinger.

In the present work, this usual, nonparametric generalised solution concept is dealt with in the context of the nowhere dense differential algebras (1.6), see subsection 7.5.2., and in particular, the commutative diagram (7.5.23), as well as Theorem G C-K, which gives a *global* version of the classical Caucy-Kovalevskaia theorem.

Then, as one of basic novelties in this work, and along the second track approach, two more concepts of solution were introduced. Namely, in (5.42), we introduced the concept of *parametric classical solution*, while in (9.1.18), the concept of *parametric generalised solution* was presented.

Needless to say, it is important to indicate the relevant *connections* between these four concepts of solutions, namely :

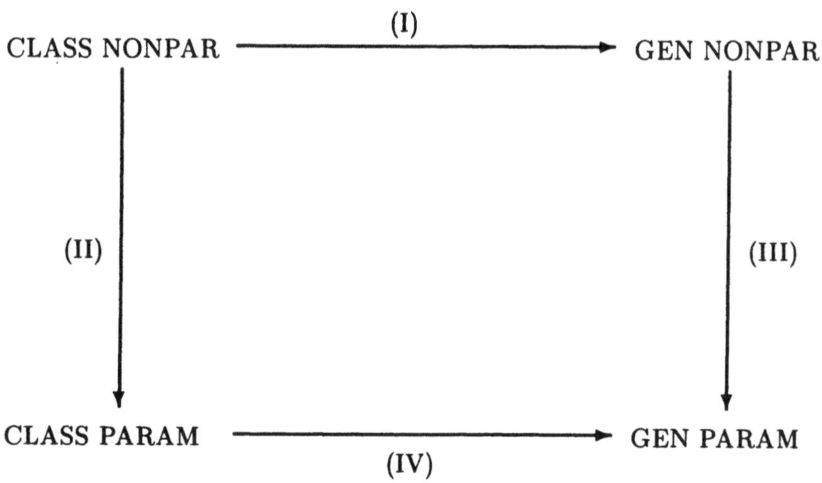

where CLASS NONPAR, GEN NONPAR, CLASS PARAM and GEN PARAM stand respectively for classical nonparametric, generalised nonparametric, classical parametric, and generalised parametric solutions.

In this regard, along the *first track*, that is, the usual nonparametric approach, the connection (I) between classical and generalised solutions is well established in the literature mentioned above. Also, in this work, certain important aspects of this connection are presented in sections 7.1., 7.3., 7.5.2., as well as later, in chapter 12.

Within the *second track*, that is, the parametric approach, aspects of the connection (IV) between classical and generalised solutions can be found earlier in this section, see for instance (9.1.14) - (9.1.17), (9.1.23).

Further, in the case of *classical* solutions, the connection (II) between their usual, nonparametric, and on the other hand, their parametric versions is presented in chapter 5, and in particular, in Theorem 1 there.

Finally, therefore, it remains to establish the connection (III) in the case of *generalised* solutions between their usual, nonparametric, and alternatively, parametric versions. This is done now in

Theorem 3

If $U \in A_\Omega$ is a usual, nonparametric generalised solution of the nonlinear PDE in (1.1), (1.2), then, see (7.3.6), (7.3.16)

$$(9.1.34) \qquad \beta_*(U) \in (\mathcal{C}^\infty(\Omega, M))^{\mathbf{N}} / \approx_{nd} \quad \subseteq \quad (\mathcal{C}^\infty_n(M))^{\mathbf{N}} / \approx_{nd}$$

is a parametric generalised solution of that equation on $D_{\beta_*(U)} = \Omega$, see (9.1.18), (9.1.21).

Proof. Let us assume that U has the representation, see (7.1.4), (7.3.7), $U = s + \mathcal{I}_{nd}(\Omega) \in A_\Omega$, where $s = (\psi_\nu | \nu \in \mathbf{N}) \in (\mathcal{C}^\infty(\Omega))^{\mathbf{N}}$. Then (7.5.24) gives $T\,U = Ts + \mathcal{I}_{nd}(\Omega) \in A_\Omega$, where $Ts = (T(x, D)\psi_\nu | \nu \in \mathbf{N})$.
Therefore, the fact that U is assumed to be a usual, nonparametric generalised solution of the PDE in (1.1), (1.4), means that, see (7.5.19), (7.5.23)

$$(T(x, D)\psi_\nu | \nu \in \mathbf{N}) \in \mathcal{I}_{nd}(\Omega)$$

which in view of (7.1.1) yields

$$\begin{aligned}
& \exists\ \Gamma \subset \Omega, \text{ closed, nowhere dense } : \\
& \forall\ x \in \Omega \setminus \Gamma\ : \\
(9.1.35) \qquad & \exists\ \Delta \subseteq \Omega \setminus \Gamma \text{ neighbourhood of } x,\ \mu \in \mathbf{N}\ : \\
& \forall\ \nu \in \mathbf{N},\ \nu \geq \mu\ : \\
& \quad T(x, D)\psi_\nu(x) = 0 \text{ on } \Delta
\end{aligned}$$

On the other hand, in view of (7.3.6), (7.3.16), it follows that

$$(9.1.36) \qquad \beta_*(U) = s_* + \mathcal{I}_{nd}(\Omega) \in A_\Omega$$

where, see (7.2.2), (3.5)

$$s_* = (\ (\psi_\nu)_* \mid \nu \in \mathbf{N}\ )$$

(9.1.37)

$$\Omega \ni x \longmapsto (\psi_\nu)_*(x) = (x, \psi_\nu(x)) \in M,\ \ \nu \in \mathbf{N}$$

Let us denote

(9.1.38) $\quad (V_\nu | \nu \in \mathbf{N}) = s_*$

then (9.1.36), (7.4.17), (7.4.18) give for $\nu \in \mathbf{N}$

(9.1.39) $\quad V_{\nu 1} = \mathrm{id}_\Omega, \quad V_{\nu 2} = \psi_\nu, \quad D_{V_\nu} = \Omega$

hence (9.1.7) results in

(9.1.40) $\quad D_{s_*} = D_{(V_\nu | \nu \in \mathbf{N})} = \Omega$

which in view of (9.1.36), (9.1.21) means that

(9.1.41) $\quad D_{\beta_*(U)} = \Omega$

But (9.1.39), (5.40), (5.41), (9.1.35), (9.1.38) imply that s_* satisfies (9.1.9), and hence (9.1.6). In this way, according to (9.1.18), (9.1.40), the proof is completed.

2. Symmetries of Generalised Solutions in A_Ω

In this section, the *advantage* of the parametric, that is, the second track approach to arbitrary nonlinear, including nonprojectable Lie group actions on usual and generalised functions will become apparent. In particular, this advantage applies as well to the Lie group symmetries of usual and generalised solutions of nonlinear PDEs of the general type in (1.1), (1.4).
This advantage becomes obvious, when we compare the general results in this section, or the more particular ones in the next section, with those obtained earlier in chapters 6 - 8.

Although the parametric approach applies equally easily to all generalised functions in the nowhere dense differential algebras A_Ω in (1.6), here, for the sake of simplicity and clarity, when we turn to applications, we shall limit ourselves to dealing with generalised functions or solutions which are elements of the subsets $S_{nd}(\Omega)$ in the nowhere dense differential algebras A_Ω, see (7.5.48).
This restriction, however, covers a large and relevant class of generalised functions and solutions, since for instance, it contains the *global* generalised solutions of arbitrary nonlinear analytic PDEs, solutions given by the extension of the classical Cauchy-Kovalevskaia theorem, see (7.5.26) - (7.5.29). Also, as seen in (7.5.48) this class goes beyond the Schwartz distributions and the Colombeau generalised functions.

Furthermore, as seen in Theorem G S below, in this section, the interest in the generalised functions in $S_{nd}(\Omega)$ goes beyond the analytic case, and reaches towards the $\mathcal{C}^\infty$-smooth case as well.

The generalised functions U in the nowhere dense differential algebras A_Ω, as much as those in their subsets $S_{nd}(\Omega)$, are given in their usual, *nonparametric* form. And we accept them in this form, in order to accommodate more easily the traditional ways of dealing with generalised functions. However, in such a form, they only serve as *input* to the parametric, that is, second track approach, and therefore, we shall immediately have to translate such usual, nonparametric generalised functions into their parametric form, see connection (III) in the diagram in section 1.

This translation, as we have seen it, is an easy matter. Moreover, we shall do it in a special form. Namely, owing to its advantages, see (3.22), (3.24), (3.25), (3.28), (3.33), (5.44), (5.47), we shall choose the *canonical* parametric form of these usual, nonparametric generalised functions, a form obtained according to, see (7.5.49) - (7.5.51), (7.3.16), (7.3.6)

(9.2.1) $$A_\Omega \ni U \longmapsto \beta_*(U) \in (\mathcal{C}^\infty(\Omega, M))^{\mathbf{N}}/\approx_{nd} \quad \subseteq \quad (\mathcal{C}_n^\infty(M))^{\mathbf{N}}/\approx_{nd}$$

Let us shortly recall here the relevant details. Once we choose any representation in the usual, nonparametric way for such a generalised function U, that is, we represent it according to the quotient structure of the nowhere dense differential algebra A_Ω, see (7.1.4), (7.3.7)

(9.2.2) $$U = s + \mathcal{I}_{nd}(\Omega) \in A_\Omega$$

where $s = (\phi_\nu | \nu \in \mathbf{N}) \in (\mathcal{C}^\infty(\Omega))^{\mathbf{N}}$, then the canonical parametric representation of U is given by, see (7.3.8), (9.2.1)

(9.2.3) $$\beta_*(U) = \text{class}_{nd}\ s_*$$

This is all what the mentioned translation of generalised functions, from their usual, nonparametric, to their parametric form will involve.

We note here that, see (3.5), (7.2.3), since $\beta_*(U)$ is obtained as a canonical parametrisation of U, its domain of parameters will be Ω, or in terms of (9.1.7), (9.1.21), we have, see also (9.1.40), (9.1.41)

(9.2.4) $$D_{\beta_*(U)} = D_{s_*} = \Omega$$

while for any other parametric representation of of $\beta_*(U)$, say

(9.2.5) $$\beta_*(U) = \text{class}_{nd}\ (V_\nu | \nu \in \mathbf{N})$$

with $(V_\nu | \nu \in \mathbf{N}) \in (\mathcal{C}^\infty(\Omega, M))^{\mathbf{N}}$, we shall have, see (9.1.12), (9.1.17)

(9.2.6) $$D_{(V_\nu | \nu \in \mathbf{N})} \supseteq \Omega \setminus \Gamma$$

where $\Gamma \subset \Omega$ is a certain closed, nowhere dense subset.

Let us now, in the particular case of $U \in S_{nd}(\Omega)$, recall shortly, see (7.5.30) - (7.5.47), that there exists

(9.2.7) $\Gamma \subset \Omega$ closed, nowhere dense, $\psi \in \mathcal{C}^\infty(\Omega \setminus \Gamma)$

such that, see (7.5.40), (7.5.42), (7.5.45)

(9.2.8) $U = \psi$ on $\Omega \setminus \Gamma$

and we can mention here once more that the interest in such generalised functions U comes, among others, from the fact that their singularity sets Γ can have arbitrary large positive Lebesgue measure, Oxtoby.

Let us now assume given a nonlinear PDE such as in (1.1), (1.4), and with it its usual Lie group of symmetry G in (1.2). Further, let us asuume that a given $U \in A_\Omega$ is a usual, nonparametric generalised solution on Ω of that equation. This means that in the sense of the commutative diagram (7.5.23), we have

(9.2.9) $T\, U = 0 \in A_\Omega$

The main purpose of this section is to apply to U, more precisely, to its canonical parametric representation $\beta_*(U)$, any of the Lie group symmetries $g \in G$, and in doing so, to make explicit the condition (9.1.33) in Theorem 2, section 9.1., which in this case becomes

(9.2.10) $D_{\beta_*(U)} \cap D_{g\, \beta_*(U)} \neq \phi$

where, see (7.3.10), (9.2.3)

(9.2.11) $g\, \beta_*(U) = \text{class}_{nd}\, g\, s_*$

First we can note now that, in view of (9.2.4), condition (9.2.10) reduces to

(9.2.12) $D_{g\, \beta_*(U)} \neq \phi$

therefore, in view of Theorems 2 and 3 in section 1, we obtain

Corollary 1

Let $U \in A_\Omega$ be a usual, nonparametric generalised solution of the nonlinear PDE in (1.1), (1.4), and let G in (1.2) be the usual Lie group of symmetry of that equation.
Given any $g \in G$, if

(9.2.13) $D_{g\, \beta_*(U)} \neq \phi$

then $g\, \beta_*(U) \in (\mathcal{C}_n^\infty(M))^{\mathbf{N}} / \approx_{nd}$ is a parametric generalised solution on the open set (9.2.13), of the PDE in (1.1), (1.4).

□

In order to make more explicit the above condition (9.2.13), we shall now restrict ourselves to usual, nonparametric generalised functions or solutions U in $S_{nd}(\Omega)$, in which case (7.5.30) - (7.5.47), and in particular, (9.2.7), (9.2.8) apply. In this way, we can obtain a more detailed information on $\beta_*(U)$ and $g\ \beta_*(U)$, and hence as well, on the open subset $D_{g\ \beta_*(U)}$ of Ω, that is, on condition (9.2.13), which is now our particular version of (9.1.33).

Recalling (7.5.42), (7.5.44), we can further specify the parametric representation (9.1.14) of U, and hence through (7.5.53), the parametric representation (9.2.11) of $g\ \beta_*(U)$. Indeed, for $g \in G$, let us denote

$$(9.2.14) \quad (V_\nu | \nu \in \mathbf{N}) = g\ s_* = (\ g\ (\psi_\nu)_*\ |\ \nu \in \mathbf{N}\ ) \in (\mathcal{C}^\infty(\Omega, M))^{\mathbf{N}}$$

then

$$\forall\ \Delta \subset\subset \Omega \setminus \Gamma,\ \Delta \text{ nonvoid, open } :$$

$$\exists\ \mu \in \mathbf{N}\ :$$

$$(9.2.15) \quad \forall\ \nu \in \mathbf{N},\ \nu \geq \mu\ :$$

$$*)\ \psi_\nu = \psi, \quad g\ (\psi_\nu)_* = g\ \psi_* \ \text{ on } \Delta$$

$$**)\ V_{\nu 1}(x) = g_1(x, \psi(x)), \quad V_{\nu 2}(x) = g_2(x, \psi(x)), \quad x \in \Delta$$

Based on (9.2.15), let us compute $D_{(V_\nu | \nu \in \mathbf{N})}$ in (9.1.7). For that, we first have to evaluate, see (7.4.18), the open sets D_{V_ν}, with $\nu \in \mathbf{N}$. However, **) in (9.2.15) gives that, see (5.8)

$$\forall\ \Delta \subset\subset \Omega \setminus \Gamma,\ \Delta \text{ nonvoid, open } :$$

$$\exists\ \mu \in \mathbf{N}\ :$$

$$\forall\ \nu \in \mathbf{N},\ \nu \geq \mu\ :$$

$$D_{V_\nu} \cap \Delta = \Delta_{(g,\psi)} \cap \Delta$$

and then (9.1.7) will obviously imply that

$$(9.2.16) \quad D_{(V_\nu | \nu \in \mathbf{N})} \supseteq \Delta_{(g,\psi)}$$

which in view of (9.2.14), (9.2.11), (9.1.21), will yield

$$(9.2.17) \quad D_{g\ \beta_*(U)} \supseteq \Delta_{(g,\psi)}$$

In this way, Corollary 1 results in

Corollary 2

Let $U \in S_{nd}(\Omega)$ with

$$(9.2.18) \quad U = \psi \ \text{ on } \Omega \setminus \Gamma$$

where $\Gamma \subset \Omega$ is closed, nowhere dense, while $\psi \in \mathcal{C}^\infty(\Omega \setminus \Gamma)$.
If U is a usual, nonparametric generalised solution of the nonlinear PDE in (1.1), (1.4), and G is the usual Lie group of symmetry of the equation, then for any $g \in G$, for which, see (5.8)

$$(9.2.19) \quad \Delta_{(g,\psi)} \neq \phi$$

$g\ \beta_*(U)$ is a parametric generalised solution on the open set $D_{g\ \beta_*(U)} \supseteq \Delta_{(g,\psi)}$ of the PDE in (1.1), (1.4).

Remark 12

As mentioned, large classes of applications can be covered by Corollary 2.
Two of such classes are described respectively by Theorem G S, below, and by its particular case in Theorem G C-K, subsection 7.5.2., which is a global version of the classical Cauchy-Kovalevskaia theorem.
The rarefaction waves and the Riemann solvers of the nonlinear shock wave equation are covered by the class of generalised solutions described in Theorem G S.
And in the next section, we shall use Corollary 2, when revisiting the issue of the nonprojectable Lie group symmetries of these generalised solutions of the nonlinear shock wave equation.

Here, we indicate shortly the essence from the point of view of applications of the condition (9.2.19) in Corollary 2. And we note that this condition is *sufficient*, while the *necessary and sufficient* condition for Corollary 2 is obviously the following

$$(9.2.20) \quad D_{g\ \beta_*(U)} \neq \phi$$

which however is less easy to make explicit.

Returning to condition (9.2.19), first we note that, in view of (7.5.47), giving

$$(9.2.21) \quad U \in \mathcal{S}_{nd}(\Omega)$$

or giving, see (9.2.18)

$$(9.2.22) \quad U = \psi \ \text{ on } \Omega \setminus \Gamma, \quad \Gamma \subset \Omega \text{ closed, nowhere dense,} \quad \psi \in \mathcal{C}^\infty(\Omega \setminus \Gamma)$$

is practically the same thing.
Therefore, as the second step, we can immediately focus on condition (9.2.19). And in order to do so, we can take any given $g \in G$, and associate with it the mapping, see (5.6), (1.2)

$$(9.2.23) \quad \Omega \setminus \Gamma \ni x \xrightarrow{\alpha} g_1(x, \psi(x)) \in \Omega$$

Further, let $J_\alpha : \Omega \setminus \Gamma \to \mathbf{R}$ be the Jacobian of α, then, see (5.8)

$$(9.2.24) \quad \Delta_{(g,\psi)} = \{ x \in \Omega \setminus \Gamma \mid J_\alpha(x) \neq 0 \}$$

Therefore, finally, we can reformulate condition (9.2.19) in the equivalent form

(9.2.25) $\exists \; x \in \Omega \setminus \Gamma \; : \; J_\alpha(x) \neq 0$

In this way, we have indeed obtained a particularly simple and useful reformulation of (9.2.19), since (9.2.22), (9.2.23), and thus J_α are easily accessible.

□

And now we present the theorem mentioned in Remark 12, see Oberguggenberger & Rosinger [p. 312]

Theorem G S

Suppose the nonlinear PDE in (1.1), (1.4)

(9.2.26) $T(x, D)U(x) = 0, \quad x \in \Omega$

has a classical solution

(9.2.27) $\psi \in \mathcal{C}^\infty(\Omega \setminus \Gamma)$, with $\Gamma \subset \Omega$ closed, nowhere dense

on the open dense set $\Omega \setminus \Gamma$.
Then, in the sense of the commutative diagrams (7.5.23), there exists a *unique global* generalised solution on the whole of Ω

(9.2.28) $U \in A_\Omega$

of the PDE in (9.2.26), such that

(9.2.29) $U = \psi \quad \text{on} \quad \Omega \setminus \Gamma$

Furthermore, we also have that

(9.2.30) $U \in S_{nd}(\Omega)$

□

We note that the difference between the above global Theorem G S and the global version of the classical Cauchy-Kovalevskaia result in Theorem G C-K, in subsection 7.5.2., is that the nonlinear PDE in (9.2.26) is only assumed to be $\mathcal{C}^\infty$ smooth, and not analytic, as in the latter theorem.

3. Revisiting the Symmetries of Shock Waves

The result in Corollary 2, in the previous section, allows a particularly *easy* treatment of the nonprojectable Lie group symmetries of the rarefaction waves and Riemann solvers of the nonlinear shock wave equation. This is indeed the case, as long as we pursue the *second track* approach, and at the end of our computations we can remain with the respective parametric results, that is, we do not try to apply to them the mapping δ in (7.4.46), in order to express these results in terms of the usual, nonparametric generalised functions in the nowhere dense differential algebras A_Ω in (1.6).

Let us now indicate the respective details.

For any *rarefaction wave* $U \in S_{nd}(\Omega) \subset A_\Omega$ in (6.5), (6.6), (8.4), we have

(9.3.1) $\quad U = \psi \quad \text{on} \quad \Omega \setminus \Gamma$

where $\Omega = (0, \infty) \times \mathbf{R}$, $\Gamma = \{ \ (t, x) \in \Omega \mid x = u_l t \ \text{ or } \ x = u_r t \ \}$, while $\psi \in \mathcal{C}^\infty(\Omega \setminus \Gamma)$, with

$$\psi(t, x) = \left| \begin{array}{lll} u_l & \text{if} & x < u_l t \\ x/t & \text{if} & u_l t < x < u_r t \\ u_r & \text{if} & u_r t < x \end{array} \right.$$

Hence, for given $u_l < u_r$, in view of Theorem S G in section 2, U is the unique generalised solution of the nonlinear shock wave equation (6.1), (8.1), which satisfies (9.3.1).

Here we shall apply the general parametric approach in sections 1 and 2 to the rarefaction wave U.
However, in order to further clarify the detailed ways such an application works, we shall not apply directly Corollary 2 in section 2. Instead, we shall pursue step by step the process in sections 1 and 2, which leads to that corollary, when applying it to the rarefaction wave U.

First we note that, since the rarefaction wave U is in its usual, nonparametric form, we have to apply to it Theorem 3 in section 1, and thus obtain

(9.3.2) $\quad \beta_*(U) \in (\mathcal{C}^\infty(\Omega, M))^{\mathbf{N}} / \approx_{nd}$

which is the parametric generalised solution, on the whole of Ω, of the nonlinear shock wave equation.

Given now any group element $\epsilon = g \in G = (\mathbf{R}, +)$ in the nonprojectable Lie group symmetry (6.9), (8.2) of the nonlinear shock wave equation, we apply Theorem 2 in section 1, to $\beta_*(U)$ in (9.3.2), and obtain that, in case

(9.3.3) $\quad D_{g \ \beta_*(U)} \neq \phi$

then

(9.3.4) $\quad g \ \beta_*(U) \in (\mathcal{C}^\infty(\Omega, M))^{\mathbf{N}} / \approx_{nd}$

is again a parametric generalised solution of the nonlinear shock wave equation, this time, on the nonvoid open subset $D_{g \ \beta_*(U)} \subseteq \Omega$.

In this way, the only problem left is to secure condition (9.3.3).
However, in view of Remark 12 in section 2, a sufficient condition for (9.3.3) is given by (9.2.25). And according to (6.17), (6.16), it is clear that (9.2.25) holds for the rarefaction wave U.

We can conclude that every nonprojectable Lie group symmetry (6.9), (8.2) of the nonlinear shock wave equation, when acting on the parametric form of the rarefaction waves, will give again a parametric generalised solution of that equation.

However, typically, such resulting generalised solutions need no longer correspond to rarefaction waves.

It is easy to see that the *Riemann solvers* (6.7), (6.8), (8.5) of the nonlinear shock wave equation can be treated in a manner similar with the above for rarefaction waves.
Furthermore, when the nonprojectable Lie group symmetries (6.9), (8.2) of the nonlinear shock wave equation are applied to Riemann solvers, in certain cases, the results obtained may be associated with Riemann solvers, see (8.66) for instance.

4. A General Approach to the Quotient Spaces $(\mathcal{C}_n^\infty(M))^{\mathbf{N}}/\approx_{nd}$

We used the relation $\approx$ in (9.1.6), (9.1.18), (9.1.19), in the special context when one of the sequences of functions involved in it is 0, and hence, when each of the terms of this particular sequence can be considered defined on the whole of Λ. Therefore, in such a situation the second problem, mentioned after (9.1.4), is avoided.

Here we shall look into the issues related to this binary relation $\approx$, and clarify them with the help of the following general construction.
First, we need a notation. Given $l \in \mathbf{N}$, $l \geq 1$, let us denote by

$$(9.4.1) \quad \mathcal{C}_n^\infty(\mathbf{R}^l)$$

the set of all the $\mathcal{C}^\infty$ smooth functions $V : \Lambda \to \mathbf{R}^l$, where $\Lambda \subseteq \mathbf{R}^n$ is any nonvoid open subset.

Suppose now given any nonvoid open subset $\Lambda \subseteq \mathbf{R}^n$. Then, *relative to* Λ, we can define the binary relation $\approx_{nd}$ on $(\mathcal{C}_n^\infty(\mathbf{R}^l))^{\mathbf{N}}$ as follows. Given any $\sigma = (V_\nu | \nu \in \mathbf{N})$, $\sigma' = (V'_\nu | \nu \in \mathbf{N}) \in (\mathcal{C}_n^\infty(\mathbf{R}^l))^{\mathbf{N}}$, with

$$(9.4.2) \quad V_\nu : \Lambda_\nu \to \mathbf{R}^l, \quad V'_\nu : \Lambda'_\nu \to \mathbf{R}^l, \quad \nu \in \mathbf{N}$$

where Λ_ν, $\Lambda'_\nu \subseteq \mathbf{R}^n$ are nonvoid, open, we denote

$$(9.4.3) \quad \sigma \approx_{nd} \sigma' \text{ on } \Lambda$$

if and only if

$$
(9.4.4)\quad
\begin{array}{l}
\exists\ \Gamma \subseteq \mathbf{R}^n,\ \text{closed nowhere dense}\ : \\
\forall\ y \in \Lambda \setminus \Gamma\ : \\
\exists\ \Lambda' \subseteq \Lambda \setminus \Gamma\ \text{neighbourhood of}\ y,\ \mu \in \mathbf{N}\ : \\
\forall\ \nu \in \mathbf{N},\ \nu \geq \mu\ : \\
\qquad *)\ \Lambda' \subseteq \Lambda_\nu \cap \Lambda'_\nu \\
\qquad **)\ V_\nu = V'_\nu\ \text{ on } \Lambda'
\end{array}
$$

It is clear that, again, as in (7.3.1) - (7.3.3), the above binary relation is *not* reflexive.

On the other hand, it is easy to see that it is both symmetric and transitive. Therefore, with the slight adaptation presented at the end of section 7.3., it can be seen as a somewhat extended version of a usual equivalence relation. Or upon preference, it can easily be turned into an *equivalence* relation on an appropriate subset of $(\mathcal{C}^\infty_n(\mathbf{R}^l))^{\mathbf{N}}$. And in view of that, in the sequel, we shall simply talk about it as such an equivalence relation.

With respect to the nonvoid open subset Λ, we note the following. Clearly, if $\Lambda' \subseteq \mathbf{R}^n$ is another nonvoid open subset, such that $\Lambda \subseteq \Lambda'$, then

$$(9.4.5)\quad \approx_{nd}\ \text{on}\ \Lambda' \quad\Longrightarrow\quad \approx_{nd}\ \text{on}\ \Lambda$$

Let us therefore, for two given sequences $\sigma = (V_\nu | \nu \in \mathbf{N})$, $\sigma' = (V'_\nu | \nu \in \mathbf{N}) \in (\mathcal{C}^\infty_n(M))^{\mathbf{N}}$, denote by

$$(9.4.6)\quad \mathcal{L}_{\sigma\ \sigma'}$$

the possibly empty set of all the open subsets Λ in $\mathbf{R}^n$, for which

$$(9.4.7)\quad \sigma \approx_{nd} \sigma'\ \text{on}\ \Lambda$$

The structure of $\mathcal{L}_{\sigma\ \sigma'}$, in case it is nonvoid, is obvious, namely, the following properties hold

$$
(9.4.8)\quad
\begin{array}{l}
*)\ \forall\ \Lambda,\ \Lambda' \in \mathcal{L}_{\sigma\ \sigma'}\ :\quad \Lambda \cup \Lambda' \in \mathcal{L}_{\sigma\ \sigma'} \\
**)\ \forall\ \Lambda,\ \Lambda' \subseteq \mathbf{R}^n\ \text{nonvoid, open}\ : \\
\qquad \Lambda \subseteq \Lambda' \in \mathcal{L}_{\sigma\ \sigma'} \quad\Longrightarrow\quad \Lambda \in \mathcal{L}_{\sigma\ \sigma'} \\
***)\ \forall\ \Lambda \in \mathcal{L}_{\sigma\ \sigma'}\ : \\
\qquad \exists\ \Gamma \subset \Lambda\ \text{closed, nowhere dense}\ : \\
\qquad \Lambda \setminus \Gamma\ \subseteq\ \liminf_{n\to\infty} (\Lambda_\nu \cap \Lambda'_\nu)
\end{array}
$$

Clearly, the above property $* * *)$ shows that, given $\sigma,\ \sigma' \in (\mathcal{C}_n^\infty(M))^{\mathbf{N}}$, the largest possible open subsets $\Lambda \in \mathcal{L}_{\sigma\ \sigma'}$ are limited by, see (9.4.2), the domains $\Lambda_\nu,\ \Lambda'_\nu \subseteq \mathbf{R}^n$ of the functions $V_\nu,\ V'_\nu$, which are terms in the respective sequences $\sigma,\ \sigma'$.
This fact justifies the definition (9.1.7), upon which, through (9.1.9), the meaning of (9.1.6) is based.

Finally, we note that the concept of equivalence defined in (9.4.3) obviously contains as particular cases the earlier similar concepts defined in (7.3.1) - (7.3.3) and at the end of section 7.6. Indeed, these particularisations are determined by the respective choices of the domains $\Lambda_\nu,\ \Lambda'_\nu$ in (9.4.2), and of the nonvoid open subsets Λ in (9.4.3).
For instance, in (7.3.1) - (7.3.3), we have the choice $\Lambda_\nu = \Lambda'_\nu = \Delta = \Lambda$.
On the other hand, at the end of section 7.6., the choice is $\Lambda_\nu = \Lambda'_\nu = \Lambda$.
Clearly, because of these particular and well defined choices, in the respective definitions (7.3.1) - (7.3.3) and at the end of section 7.6., there was no need for a further explicit specification of the nonvoid open subsets Λ.
And it is precisely for that reason, namely, of *not* having to specify explicitely the subsets Λ, that we have chosen to work with the mentioned particular versions of the general definition in (9.4.3). Indeed, as seen in (9.4.8), and even in (9.1.9), (9.1.7), (9.1.6), the specification of these subsets Λ is a rather involved matter, when we use the general definition of the equivalence relation in (9.4.3).

5. Conclusions

The advantages of the parametric approach to arbitrary nonlinear, including nonprojectable Lie group actions on generalised functions, as well as of its applications to the Lie group symmetries acting on classical or generalised solutions, as presented in sections 1 - 3, are by now quite obvious. These advantages are further shown, when we compare the parametric approach with the usual, nonparametric one, such as in chapters 6 - 8.
And it is important to point out that the parametric approach is *sufficient* in itself, unless at the end of our computations there is a need to express explicitely in usual, nonparametric form the results of Lie group actions on classical or generalised functions, or in particular, the results of Lie group symmetries on classical or generalised solutions.

Among these advantages are the following.
Arbitrary Lie group actions can be performed in a natural, easy way, and without any restrictions, owing to the functorial advantage in the definition (4.1) - (4.4). Furthermore, these actions can be performed globally, and the results obtained are also defined globally.
And when the Lie group actions correspond to symmetries of solutions of PDEs, the conditions required are not related to the performance of such actions, which as in the general case, can be performed without restrictions. Indeed, the respective conditions are only related to securing that the results of Lie group symmetries

acting on solutions will again be solutions. And these conditions only refer to the avoidance of the singular points of the parametrisations of the solutions involved. In other words, when we use the parametric method, the onus is on us to avoid ill conceived parametrisations which may burden us with too many singular points, points that could have been avoided by better parametrisations.
In this way, the respective conditions, when performing through the parametric approach Lie group symmetries on solutions of PDEs, are natural and minimal.

Here it should be mentioned that in the case of arbitrary Lie group actions on generalised functions, the advantages of the parametric approach rely to a good extent on the *special quotient structure* of the nowhere dense differential algebras (1.6), respectively, of the larger spaces $(\mathcal{C}^\infty(\Omega, M))^{\mathbf{N}} / \approx_{nd}$, $(\mathcal{C}^\infty_{par}(\Omega))^{\mathbf{N}} / \approx_{nd}$, or $(\mathcal{C}^\infty_n(\Omega))^{\mathbf{N}} / \approx_{nd}$.
Indeed, it may not come as a surprise to see that similar advantages of the parametric approach to Lie group actions would not so easily be reproduced, when these actions are performed on other types of spaces of generalised functions.

Finally, as far as the idea of using the parametric approach to Lie group actions is concerned, one can note the following.

Whenever Lie group symmetries (1.2), (1.3), (1.5) are associated with PDEs in (1.1), (1.4), the respective Lie group actions do *not* in general make any distinction between the independent and dependent variables, since they act simultaneously on both of them, that is, in the mentioned notation, they act on the sets M. In this way, in the general case, the Lie groups of symmetry of solutions of PDEs are from the start intrinsically *compatible* with the parametric approach.
And in fact, one may even go as far as to say that they *require* the parametric approach.
The appreciation of this rather obvious and simple fact was actually at the root of the introduction in this work of the parametric approach to Lie group actions on classical and generalised functions.

With this chapter, the presentation in this work of the main features of the parametric approach is completed.
The next four chapters, that is, 10 - 13, will deal with issues which to a certain extent are related to the parametric approach.

Here it should be mentioned that the last chapter, that is, 13, occupies a *special* place, as it presents a first introduction to the theory of *genuine* Lie semigroup actions. In other words, it introduces actions by Lie semigroups which are *not* subsemigroups of any Lie group.
As was already mentioned, the study of such genuine Lie semigroups has not been developed earlier, since the Lie semigroups considered so far in the literature, see for instance Hilgert & Neeb, are assumed to be subsemigroups of Lie groups.

On the other hand, as we shall see in chapter 13, the interest in genuine Lie semigroups can be significant, since they can considerably extend the classes of transformations - called in the sequel *semisymmetries* - which turn solutions of nonlinear PDEs into other solutions. Indeed, as the examples in chapter 13 show, it will now be possible to use transformations of M in (1.2), (1.3), (1.5), which need no longer be diffeomorphisms of M, and thus invertible, but can be arbitrary $\mathcal{C}^\infty$-smooth mappings of M.

It can sometime happen that the merit in the novelty of a work, no matter how relevant or otherwise it may be in itself, may in fact prove to be somewhat secondary to certain possible openings to which the work happens to point, openings which however need not so clearly belong to the main lines intended and pursued by the work itself.
And it appears that, indeed, such may be the case with the opening in chpater 13.
In order to further clarify the extent of the step taken forward, when going from Lie group actions, to actions by genuine Lie semigroups, let us mention in the case of a nontrivial manifold M, how much *larger* is the semigroup of $\mathcal{C}^\infty$-smooth mappings of M to M, than the group of $\mathcal{C}^\infty$-smooth diffeomorphisms of M. And obviously, that former semigroup *cannot* be a subsemigroup of any group.

Chapter 10

Projectable Lie Group Actions and Hilbert's Fifth Problem

Here, in order to facilitate the presentation in chapter 11, we give, mostly without proofs, a short account of some earlier, related results, see for details Oberguggenberger & Rosinger, Rosinger & Rudolph, and Rosinger & Walus [1,2].
Specifically, we shall shortly recall the way projectable Lie group actions were extended to the nowhere dense differential algebras of generalised functions A_Ω, see (1.6), (7.1.4), (7.1.3).
In the research monograph and three papers mentioned, projectable Lie group actions were also extended to the Colombeau algebras, as well as to the solutions given by the new method for solving arbitrary continuous nonlinear PDEs, method based on the order completion of spaces of continuous functions, and introduced in Oberguggenberger & Rosinger.
Here however, owing to the fact that the present work concentrates on the Lie group actions on nowhere dense differential algebras, we shall only deal with these algebras.

In the case of the nowhere dense differential algebras, this amounts to a further contribution to the solution of Hilbert's fifth problem, when this problem is considered in its *fully general* original formulation, that is, of Lie group actions on outside entities - manifolds or functions, for instance - and not only in the more narrow sense of the intrinsic Lie group structure, independent of any outside actions, see for more details the beginning of chapter 11.

In chapter 11, based on the results in chapters 5 - 9, we shall give a yet more comprehensive solution to the general version of Hilbert's fifth problem. Indeed, in this work, arbitrary nonlinear, including nonprojectable Lie group actions were defined on the nowhere dense differential algebras of generalised functions A_Ω, see

(1.6), (7.1.4), (7.1.3). And these algebras include the generalised functions from $S_{nd}(\Omega)$, see (7.5.48), which can have rather *large and arbitrary singularities*, such as those on any closed, nowhere dense subsets $\Gamma \subset \Omega$, subsets which, as is known, can be of arbitrary large positive Lebesgue measure.
It follows that the very possibility of constructing arbitrary Lie group actions on entities with such sort of significantly large singularities does indeed give a rather comprehensive answer to Hilbert's fifth problem, when this problem is considered in its full generality.
Namely, it shows that *smoothness* conditions need *not* at all be involved, even when, as in this work, the most general type of nonlinear and nonprojectable Lie group actions are constructed.

Now we return in this chapter to the short presentation of the mentioned earlier results in Oberguggenberger & Rosinger. Let us therefore consider that the Lie group action in the framework of (1.1) - (1.5) is *projectable*, that is, we have, see (2.3)

$$(10.1) \quad g_1(x,u) = g_1(x), \quad g \in G, \ (x,u) \in M$$

In this case, for each $g \in G$, the $\mathcal{C}^\infty$-smooth diffeomorphism, see (1.5), (1.2)

$$(10.2) \quad M \ni (x,u) \xrightarrow{g} g(x,u) = (g_1(x), g_2(x,u)) \in M$$

has the component, see (2.4)

$$(10.3) \quad \Omega \ni x \xrightarrow{g_1} g_1(x) \in \Omega$$

itself a $\mathcal{C}^\infty$-smooth diffeomorphism.

This is precisely the crucial simplifying fact which makes projectable Lie group actions so much simpler and easier to handle, including when they act on the nowhere dense differential algebras of generalised functions.
On the other hand, projectable Lie group actions are not trivial, since they contain the coordinate tranforms used in General Relativity.
One of the main limitations of projectable Lie group actions, however, comes from the fact that important nonlinear PDEs have the symmetry groups of their solutions given by nonprojectable Lie group actions, even in such simple cases as, for instance, that of the nonlinear shock wave equation, see chapters 6, 8 and 9.

As a start, let us recall the way projectable Lie groups act on usual functions. Let $U \in \mathcal{C}^\infty_{par}(\Omega)$, which means that $U : \Delta \to \mathbf{R}$, for a certain nonvoid open subset $\Delta \subseteq \Omega$. In this case, (1.2), (1.3), (10.1) give for a projectable Lie group action $g \in G$

$$(10.4) \quad \begin{aligned} \bar{x} &= g_1(x), \quad x \in \Delta \\ \bar{u} &= g_2(x,u), \quad x \in \Delta, \ u \in \mathbf{R} \end{aligned}$$

therefore, the natural way, see (2.8) - (2.10), to define the group action $\bar{U} = gU : \bar{\Delta} \to \mathbf{R}$ of g on U, is by

$$\bar{\Delta} = g_1(\Delta)$$

$$(10.5) \qquad \bar{x} = g_1(x), \quad x \in \Delta$$

$$\bar{U}(\bar{x}) = g_2(x, U(x)), \quad x \in \Delta$$

and in view of (10.3), this can indeed be done, simply by

$$(10.6) \qquad \bar{U}(\bar{x}) = g_2(g_1^{-1}(\bar{x}), U(g_1^{-1}(\bar{x})), \quad \bar{x} \in \bar{\Delta}$$

As was first shown in Oberguggenberger & Rosinger [pp. 299-353], this simplicity of projectable Lie group actions on usual functions will as well transfer to actions on the nowhere dense differential algebras of generalised functions A_Ω in (1.6). And for convenience, we shall shortly recall here this result.
Namely, for a given projectable Lie group action (1.2), (1.3), (10.1), we shall show an easy and direct way to lift that action on M, to an action

$$(10.7) \qquad G \times A_\Omega \longrightarrow A_\Omega$$

where we use the following form of the nowhere dense differential algebras A_Ω, see (7.1.4), (7.1.3)

$$(10.8) \qquad A_\Omega = (\mathcal{C}^\infty(\Omega))^{\mathbf{N}} / \mathcal{I}_{nd}(\Omega)$$

This time, owing to the property (10.1) of projectable Lie group actions, we can avoid all the difficulties we encountered in sections 7.2. - 7.6., and can define the action (10.7) in a straightforward manner, even within the first track, usual, nonparametric approach. Indeed, in view of (10.6), given $g \in G$, we can define

$$(10.9) \qquad (\mathcal{C}^\infty(\Omega))^{\mathbf{N}} \ni s = (\psi_\nu | \nu \in \mathbf{N}) \longmapsto gs = (g\psi_\nu | \nu \in \mathbf{N}) \in (\mathcal{C}^\infty(\Omega))^{\mathbf{N}}$$

and note that, owing to an obvious and convenient consequence of the asymptotic vanishing property (7.1.1) of the nowhere dense ideal $\mathcal{I}_{nd}(\Omega)$, we have the important relation, see Oberguggenberger & Rosinger [p. 317]

$$(10.10) \qquad \begin{array}{l} \forall \ g \in G, \ s \in (\mathcal{C}^\infty(\Omega))^{\mathbf{N}}, \ w \in \mathcal{I}_{nd}(\Omega) \ : \\ \quad g(s+w) - gs \in \mathcal{I}_{nd}(\Omega) \end{array}$$

Clearly, the relations (10.9), (10.10) mean that the actions (10.9) are compatible with the quotient structure in (10.8), therefore, we have indeed managed to define the projectable Lie group actions (10.7).

Needless to say, the above construction in (10.1) - (10.6), (10.8) - (10.10), of the projectable Lie group actions (10.7) on the nowhere dense differential algebras of generalised functions A_Ω in (10.8), is indeed very simple, when compared with the constructions in chapter 7, aimed to define arbitrary nonlinear, including nonprojectable Lie group actions on the same algebras.

In this respect, regardless of its possible advantages, the second track, parametric approach to generalised functions, introduced in this work, may also be seen as

the attempt to pass over the significant threshold from the nonprojectable to the projectable Lie group actions, and do so in as simple as possible, even if perhaps somewhat unusual and strange, way.
The fact, however, is that although in the case of projectable Lie group actions one could easily stay within the first track, usual, nonparametric approach, and do so both in the case of classical and generalised functions, the latter being elements in the algebras A_Ω, on the other hand, once we have embarked upon the idea of the second track, parametric approach in dealing with nonprojectable Lie group actions, now, after having developed the present work, this parametric approach seems to be the most natural one, even in the case of nonprojectable Lie group actions on classical functions.
And in particular, the results in Theorem 1, in chpater 5, and Theorems 2, 3, and Corollaries 1, 2, in chapter 9, seem to support such a perception.

In order to complete the picture about the natural and easy way projectable Lie groups act on the nowhere dense differential algebras, and can do so within the first track, usual, nonparametric approach, we recall the respective result in Oberguggenberger & Rosinger, on the symmetries of generalised solutions of nonlinear PDEs. This result is the earlier, particular case of the respective extensions to arbitrary nonlinear, including nonprojectable Lie group symmetries of generalised solutions, obtained in this work in the mentioned Theorems 2, 3, and Corollaries 1, 2, in chapter 9. Further details on this result can be found in Oberguggenberger & Rosinger [pp. 321-325].

In order to avoid unnecessary technical complications, we shall stay with the formulation of the concept of generalised solutions presented in subsection 7.5.2., and in particular, in the commutative diagram (7.5.23), see also the context of Theorem G S, in section 9.2.
Suppose given, as in (1.1), (1.4), a nonlinear PDE

$$(10.11) \quad T(x,D)U(x) = 0, \quad x \in \Omega$$

Then similar with (10.10), we have

$$(10.12) \quad \begin{array}{l} \forall\ s \in (\mathcal{C}^\infty(\Omega))^{\mathbf{N}},\ w \in \mathcal{I}_{nd}(\Omega)\ : \\ \\ T(x,D)(s+w) - T(x,D)s \in \mathcal{I}_{nd}(\Omega) \end{array}$$

where for $t = (\chi_\nu | \nu \in \mathbf{N}) \in (\mathcal{C}^\infty(\Omega))^{\mathbf{N}}$, we denote $T(x,D)t = (T(x,D)\chi_\nu | \nu \in \mathbf{N})$. It follows that in view of (10.12), we can obviously define

$$(10.13) \quad T : A_\Omega \longrightarrow A_\Omega$$

where for $U = s + \mathcal{I}_{nd}(\Omega) \in A_\Omega$, we put $TU = T(x,D)s + \mathcal{I}_{nd}(\Omega)$.

Now the mentioned main result can be formulated as follows.
Suppose G is a projectable Lie subgroup of the usual Lie group of symmetry of the nonlinear PDE in (10.11). Let $U \in A_\Omega$ be any global generalised solution of, see (10.13)

(10.14) $\quad T\,U = 0$

Then for every $g \in G$, we shall have, see (10.7)

(10.15) $\quad T\,(gU) = 0$

that is, the projectable Lie group action of $g \in G$ on the generalised solution $U \in A_\Omega$ of (10.4), that is, $gU \in A_\Omega$, will again be a generalised solution of(10.4).

Since the proof of this earlier important result is particularly simple, and also instructive with respect to the properties of the nowhere dense differential algebras of generalised functions A_Ω, we shall reproduce it here.
In view of (10.8), we can assume that U in (10.14) is given by

(10.16) $\quad U = s + \mathcal{I}_{nd}(\Omega) \in A_\Omega, \quad s = (\psi_\nu | \nu \in \mathbf{N}) \in (\mathcal{C}^\infty(\Omega))^{\mathbf{N}}$

then (10.9), (10.10) result in

(10.17) $\quad gU = gs + \mathcal{I}_{nd}(\Omega) \in A_\Omega, \quad gs = (g\psi_\nu | \nu \in \mathbf{N}) \in (\mathcal{C}^\infty(\Omega))^{\mathbf{N}}$

Thus according to (10.13), the relation (10.15) is equivalent with

(10.18) $\quad (T(x, D)(g\psi_\nu) | \nu \in \mathbf{N}) \in \mathcal{I}_{nd}(\Omega)$

However, (10.14), (10.16), (10.13) yield

(10.19) $\quad (T(x, D)\psi_\nu | \nu \in \mathbf{N}) \in \mathcal{I}_{nd}(\Omega)$

hence (7.1.1) implies that

$$\exists\ \Gamma \subset \Omega \ \text{ closed, nowhere dense } :$$

$$\forall\ x \in \Omega \setminus \Gamma \ :$$

(10.20) $$\exists\ \Delta \subseteq \Omega \setminus \Gamma \ \text{ neighbourhood of } x, \ \ \mu \in \mathbf{N} \ :$$

$$\forall\ \nu \in \mathbf{N}, \ \nu \geq \mu \ :$$

$$T(y, D)\psi_\nu(y) = 0, \quad y \in \Delta$$

In other words, for large enough ν, each ψ_ν is a *classical* solutions of (10.11) on a neighbourhood of each point in $\Omega \setminus \Gamma$. Therefore, for $\nu \in \mathbf{N}$, we consider the open subsets in $\Omega \setminus \Gamma$, given by

(10.21) $$\Omega_\nu = \left\{ x \in \Omega \setminus \Gamma \ \middle| \ \begin{array}{l} \exists\ \Delta \subseteq \Omega \setminus \Gamma \ \text{ neighbourhood of } x \ : \\ T(y, D)\psi_\nu(y) = 0, \quad y \in \Delta \end{array} \right\}$$

Then clearly, for $\nu \in \mathbf{N}$, we have

(10.22) $\quad T(x, D)\psi_\nu(x) = 0, \quad x \in \Omega_\nu$

On the other hand, in view of (7.1.1), it follows that (10.20) is equivalent with

$$(10.23) \quad \bigcup_{\mu \in \mathbf{N}} \bigcap_{\nu \in \mathbf{N},\, \nu \geq \mu} \Omega_\nu = \Omega \setminus \Gamma$$

Indeed, (10.20) $\Longrightarrow$ (10.23), since the inclusion '$\subseteq$' in (10.23) is a direct result of (10.21), while the inclusion '$\supseteq$' in the same (10.23) follows clearly from (10.20). On the other hand, also (10.23) $\Longrightarrow$ (10.20), since for any $x \in \Omega \setminus \Gamma$, the relations (10.23), (10.21) obviously imply that

$$\exists \; \mu \in \mathbf{N} \; :$$

$$\forall \; \nu \in \mathbf{N}, \; \nu \geq \mu \; :$$

$$T(x, D)\psi_\nu(x) = 0$$

and then based on a Baire category argument, see Rosinger [6, pp. 19-22], Rosinger [7, pp. 248-253], we obtain as well the stronger property in (10.20).

Given now any $g \in G$, in view of (10.3) it follows that

$$(10.24) \quad \Sigma = g_1(\Gamma) \subset \Omega \quad \text{is closed and nowhere dense}$$

and thus

$$(10.25) \quad \Omega \setminus \Gamma \xrightarrow{g_1} \Omega \setminus \Sigma$$

is a $\mathcal{C}^\infty$-smooth diffeomorphism of nonvoid open subsets of Ω. In this way, (10.23) - (10.25) result in

$$(10.26) \quad \bigcup_{\mu \in \mathbf{N}} \bigcap_{\nu \in \mathbf{N},\, \nu \geq \mu} g_1(\Omega_\nu) = \Omega \setminus \Gamma$$

However, we assumed that G is a group of usual Lie symmetries of the PDE in (10.11), therefore, (10.22), (10.25) imply that

$$(10.27) \quad T(x, D)\psi_\nu = 0 \quad \text{on} \quad g_1(\Omega_\nu) \subseteq \Omega \setminus \Sigma$$

But then (10.26), (10.27) give

$$\forall \; x \in \Omega \setminus \Sigma \; :$$

$$\exists \; \mu \in \mathbf{N} \; :$$

$$\forall \; \nu \in \mathbf{N}, \; \nu \geq \mu \; :$$

$$T(x, D)\psi_\nu(x) = 0$$

which as above with the equivalence between (10.20) and (10.23), will finally result in

$$(T(x, D)(g\psi_\nu) \,|\, \nu \in \mathbf{N}) \in \mathcal{I}_{nd}(\Omega)$$

and the proof of (10.15) is completed.

Chapter 11

Nonprojectable Lie Group Actions and an Answer to Hilbert's Fifth Problem

It appears that, in the more recent literature, there is less than complete clarity or agreement about the original meaning of Hilbert's fifth problem. This may not be so surprising, since the fifth problem happens to be one the few of those twenty three, formulated by Hilbert in 1900, which, although of a clear mathematical intent and content, nevertheless, has a certain rather inevitable vagueness or ambiguity, owing to the nature of the problem itself.
Let us therefore recall certain relevant passages from Hilbert's original formulation of his fifth problem, as well as a few more recent and related comments, see Hilbert, Browder, Yang, Hirschfeld, Aczel, Olver [3].

Two passages from Hilbert are particularly relevant, see Hilbert, or Browder [pp. 12-14], namely :

> 'How far Lie's concept of continuous groups of transformations is approachable in our investigations without the assumption of the differentiability of the functions ?'

and

> 'In how far are the assertions which we can make in the case of differentiable functions true under proper modifications without this assumption ?'

In order to illustrate the mentioned insufficient clarity and agreement about the interpretation of Hilbert's fifth problem, let us note here that, in Aczel, for instance, the meaning of the second citation above is focused solely on functional equations, like the well known equation of Abel

$$f(z+w) = f(z)f(w), \quad z,\ w \in \mathbf{C}$$

And it is considered there that the second part of Hilbert's fifth problem is limited to asking whether the smoothness conditions involved in such functional equations can be weakened, when we try to solve them.

Usually, see Yang, Hirschfeld, it is considered that the main part of Hilbert's fifth problem, related to the first citation above, was solved, and given a positive answer, in the 1952 result of Montgomery, Zippin and Gleason, namely :

Theorem M-Z-G

Every locally Euclidean group is a Lie group.

□

However, in Olver [3], it is pointed out that this result only addresses a special case of action in Hilbert's fifth problem, namely, when the group acts only upon itself.
This comment is particularly relevant in the context of (1.1) - (1.5), for instance, when we deal with Lie group symmetries of solution of PDEs. Indeed, such symmetries can hardly be treated as the mentioned kind of special actions of a Lie group, that is, actions which are merely upon the group itself.
Furthermore, in his own formulation of the fifth problem, Hilbert specifically writes down the kind of Lie group actions he considers, namely

$$(11.1) \quad x'_i = f_i(x_1, \ldots, x_n; a_1, \ldots, a_r), \quad i = 1, \ldots, n$$

where quite clearly, $x_1, \ldots, x_n$ and similarly $x'_1, \ldots, x'_n$ are the coordinates on a set which, typically, is outside of, and different from the group, a set upon which the Lie group is supposed to act, while $a_1, \ldots, a_r$ are the group parameters.
And the fact is that there has been an awareness about the more general, and appropriate meaning of Hilbert's fifth problem, namely, related to the situations when the Lie groups act not only on themselves, but also on manifolds, functions, and possibly other more involved entities as well. Indeed, as reviewed in Yang, for instance, there have been results in this direction, starting with the 1940s. One of such results is attributed to Bochner, Montgomery and Kuranishi, and it goes as follows

Theorem B-M-K

Let G be a locally compact group which acts effectively on a differentiable manifold M, such that, for every $g \in G$, the action $M \ni x \longmapsto gx \in M$ is a differentiable mapping of M into itself. Then G is a Lie group and (G, M) is a differentiable transformation group.

□

In this work, we shall take the more general view of Hilbert's fifth problem, namely, when the Lie groups act on outside entities which can be as removed or different from usual manifolds, as for instance, the nowhere dense differential algebras of generalised functions.

In this sense, Hilbert's fifth problem can be seen as asking whether Lie group actions could be defined on entities with such a highly *deficient* degree of smoothness as the generalised functions.
And as seen in chapters 6 - 9, this question does have a relevance of its own, which is independent of the fact that, back in 1900, Hilbert asked a question that may be construed as related to it. Indeed, it is nowadays well known that important nonlinear PDEs have relevant solutions which fail to be classical. Furthermore, there is no known a priori argument why the Lie group symmetries of such equations should only be considered in the context of classical solutions, and why should relevant generalised solutions be prevented from the actions of such symmetries.

The reason, therefore, that we associate certain results in this work with Hilbert's fifth problem is, simply, that this problem . . . has been there for quite a long time by now . . .
Furthermore, by typically being about Lie group actions on entities outside of, and different from the Lie groups themselves, and its question being about the *minimal* necessary smoothness conditions involved, this fifth problem of Hilbert cannot but quite naturally - and all on its own and by itself - be connected with the issue of Lie group actions on generalised functions.
Also, as mentioned, the existing less than settled points of view on the proper meaning of Hilbert's fifth problem may benefit from a further pointing out of the fact that, in his original formulation, Hilbert was concerned with the general situation of Lie groups acting not only on the group themselves, but on outside and different entities as well.

In this context of seeing Hilbert's fifth problem, let us now recall in short the main results in some of the earlier chapters which show that arbitrary nonlinear, and in particular, nonprojectable Lie group actions can be defined on such *highly nonsmooth* entities as the nowhere dense differential algebras of generalised functions, as well as on some of their still more general extensions.

As before, let us start with the situation in (1.2), (1.3), namely, we are given a nonvoid open subset $\Omega \subseteq \mathbf{R}^n$, and we define $M = \Omega \times \mathbf{R}$. Then on M, we consider the arbitrary Lie group action

(11.2) $\quad G \times M \ni (g, (x, u)) \longmapsto g(x, u) \in M$

where $g(x, u) = (g_1(x, u), g_2(x, u))$, while $g_1(x, u) \in \Omega, \quad g_2(x, u) \in \mathbf{R}$.

As seen in (10.1), (10.7), in the particular case of projectable Lie group actions (11.2), one can easily extend such an actions to the nowhere dense differential algebras of generalised functions A_Ω, namely

(11.3) $\quad G \times A_\Omega \longrightarrow A_\Omega$

and this result has been presented in full detail in Oberguggenberger & Rosinger, Rosinger & Rudolph, and Rosinger & Walus [1,2].

In the case of arbitrary Lie group actions (11.2), the initial aim of chapter 7 was to obtain the corresponding extended version of Lie group actions in (11.3).

This, however, led to a deeper analysis of the issues involved, and we found out that in order to obtain (11.3) for arbitrary Lie group actions, it was particularly convenient to turn to the *parametric* representation of the generalised functions in A_Ω. The respective stages of analysis went through (7.3.1), (7.3.12) - (7.3.17), (7.4.64), and we arrived to the commutative diagrams (7.4.65), (7.4.66), (7.6.3) and (7.6.11). Finally, in chpater 9, we concluded with the general parametric approach, which replaced the nowhere dense differential algebras of generalised functions A_Ω in (11.2) with still more general spaces. Indeed, first, we arrived at the spaces, see (7.13.17), $(\mathcal{C}^\infty_{par}(\Omega))^{\mathbf{N}} / \approx_{nd}$, which are but a collection of nowhere dense differential algebras A_Δ, for all nonvoid open subsets $\Delta \subseteq \Omega$. Then we had to consider the *paramteric* form of such spaces, namely $(\mathcal{C}^\infty_{par}(\Omega, M))^{\mathbf{N}} / \approx_{nd}$, which are no longer algebras, see also (7.4.65), (7.6.11), (7.6.15).
Finally, in (9.1), (9.4.1) - (9.4.8), we arrived at the still more general *parametric* spaces of generalised functions $(\mathcal{C}^\infty_n(M))^{\mathbf{N}} / \approx_{nd}$.
And it was precisely with these spaces of parametric generalised functions that we managed to obtain the extension of (11.3), valid this time for arbitrary nonlinear, and in particular, nonprojectable Lie group actions, namely

$$(11.4) \quad G \times ((\mathcal{C}^\infty_n(M))^{\mathbf{N}} / \approx_{nd}) \longrightarrow (\mathcal{C}^\infty_n(M))^{\mathbf{N}} / \approx_{nd}$$

In the above mentioned context of Hilbert's fifth problem, this result in (11.4) means that arbitrary Lie group actions can now be defined on entities which are at least as *nonsmooth* as those in the algebras of generalised functions A_Ω, and at the same time, range over yet *larger* spaces.
Furthermore, in view of the results in chapter 9, such an extension of Lie group actions proves to have relevance in solving large classes of nonlinear PDEs, and in particular, in studying their generalised solutions.

Chapter 12

Singularities and the Nowhere Dense Algebras of Generalised Functions

Let us present certain explicit details about why in this work, we use the nowhere dense differential algebras (1.6), (7.1.4), (7.1.3). We shall do so along two lines :

First, we shall shortly recall the very *reasons*, as well as the way the nowhere dense differential algebras were first constructed and also found useful, back in the mid 1970s, see Rosinger [3-5].
Then, we shall further elaborate on the recent interest in their use in dealing with large classes of singularities in abstract differential geometry, and in particular, de Rham cohomology and general relativity.

The question about the role of the nowhere dense differential algebras - regardless of the fact that they were the first to be used in any systematic nonlinear theory of generalised functions - is indeed a question which may quite legitimately arise, in view of the following two facts :

First, as mentioned earlier, in Rosinger [6], a wide ranging purely algebraic *characterization* was given for all possible differential algebras of generalised functions which contain the Schwartz distributions. And this characterization, together with several examples in Rosinger [6], show that there are *infinitely many* rather different *classes* of such differential algebras. Moreover, based solely on their respective algebraic constructions, there is no a priori way to determine among them certain canonical algebras. Indeed, when seen from a purely theoretical, algebraic point of view, it is clear that one cannot choose a distinguished or privileged algebra, or class of such algebras. This phenomenon is directly related to the quotient structure of such algebras, and it depends on a very simple algebraic, more precisely, ring theoretic fact, which shows that the ideals used in such quotient constructions do not, in purely algebraic terms, allow us to distinguish any particular one among them, see Rosinger [6, pp. 118,119], [8, pp. 98,99]. This situation is quite similar with, and in fact it is in a way an extension of, the well known fact that, when constructing various instances of the nonstandard reals $^*\mathbf{R}$, there is no canonical way for choosing the respective ultra filters.

Second, the later introduced, Colombeau type algebras, one of the types included in the mentioned algebraic characterization, see Rosinger [7], have recently known a certain popularity in a number of applications developed by a variety of researchers, and related to generalised or numerical solutions of nonlinear PDEs, stochastic PDEs, or general relativity, see the earlier mentioned literature, while for stochastic applications, see Albeverio, et., al., Oberguggenberger [2], Oberguggenberger & Russo, Russo, Martias [1,2], Lozanov Crvenkovic & Pilipovic.

As mentioned, in order to answer the above question related to our choice of the nowhere dense differential algebras of generalised functions, we shall give certain details about the original reasons underlying their construction. Then as well, we shall further indicate some of their recent uses which go beyond the initial range of applications developed in relation to solving nonlinear PDEs, applications which, this time, reach as far as recent joint work with A Mallios, on de Rham cohomology in abstract differential geometry, see Mallios & Rosinger.

And now, to the original reasons for constructing the nowhere dense differential algebras of generalised functions.

Generalised Functions Means Dealing With Singularities. It is rather instructive to see how the issue of singularities of functions happened to become particularised, and thus to some extent marginalised, in various theories of generalised functions, although quite clearly, it had at the beginning been the major reason for the emergence of these theories.

Indeed, the Sobolev and Schwartz linear version of the theory of generalised functions was, as is well known, developed in order to deal with nonclassical solutions of linear PDEs. And nonclassical, in this context, means solutions which are not smooth enough, that is, their order of smoothness is less than the order of partial differentiations in the respective PDEs.
In this way, a new concept of *singularity* was introduced and dealt with in the study of solutions of PDEs, namely, given by the subsets Γ of the domain Ω of the respective PDEs, subsets on which the solutions failed to be smooth enough. Indeed, the earlier concept of singularity was typically related only to such points or subsets in the neighbourhood of which the solutions became unbounded. One notable exception to this classical view of singularities was that of the Riemann solvers of the nonlinear shock wave equation, which proved to point early towards the new concept of singularity.

The interest in the new type of singularity first appeared explicitely in the late 1920s, within a simple linear framework, in the case of the so called Dirac delta function δ, although earlier, towards the end of the 1800s, Heaviside had encountered a similar phenomenon in his operational calculus aimed at solving constant coefficient ODEs.

Quite amusingly, in the linear theory of Schwartz distributions, a great stress is placed from the very beginning on the fact that the respective approach can, indeed, give a rigorous and general treatment of the Dirac delta function, which now is called the Dirac delta *distribution*, owing to the fact that it is quite directly related to a certain atomic measure.

However, the singularity of the Dirac delta distribution is of a *doubly* particular nature. Indeed, in the one dimensional case, for instance, it is concentrated at one point only, that is, $\Gamma = \{ x_0 \} \subset \Omega = \mathbf{R}$. Furthermore, the singularity of the Dirac delta distribution is nothing else but that of an atomic measure. In other words, in the terms of Schwartz distributions, it is a second order distribution, being the second derivative of a continuous function.
Certainly, owing to the power of the Lebesgue integral, the Schwartz distributions can also deal with the singularities of any locally integrable function. However, in view of Lusin's classical theorem on the structure of arbitrary measurable functions, these kind of singularities are measure wise as near as one wants to those of continuous functions.

On the other hand, when trying to deal with the classical singularities, that is, of smooth functions which become unbounded in the neighbourhood of a point, or of some subset Γ in Ω, the approach of the Schwartz distributions starts to hit *restrictions* related to the *rate of growth* of the smooth functions involved, around their classical singularities. This sort of phenomenon is well illustrated by the manner in which the approach in Schwartz distributions, for instance, can - or for that matter, it cannot - define the so called principal values of certain smooth functions which are singular at one single point in Ω, and at the same time also fail to be locally integrable in the neighbourhood of that point.

However, such a limitation of the usual Schwartz and Sobolev distributional approach is not at all insignificant. Indeed, the issue of such point, or more involved, classical singularities is most relevant, even in the particular case of *analytic* functions. And with analytic functions of several complex variables, we do not yet know enough about the structure of their singularity sets, or about their rate of growth near to singularities.

The more modern mathematical expression of this limitation of the linear distribution theory is that the Schwartz distributions fail to be a *flabby* sheaf. In other words, a ditribution T_0 on a nonvoid open subset Ω_0 of Ω cannot in general be extended to a distribution T on the whole of Ω.

All in all, the linear distribution theory of Schwartz is limited at least in three ways :

- It is linear only.

- It only allows for singularities which are on subsets of Lebesgue measure zero.

- It requires certain growth conditions around singularities.

However, regardless of the above significant limitations on their capabilities to handle sigularities, the various generalised functions theories have not yet tried to address them, and go beyond them.

Needless to say, this failure becomes even more questionable, when we recall that, since the 1970s, studies in fractal sets and strange attractors have progressed significantly. Moroever, their applications to the study of nonlinear PDEs is becoming more and more relevant, specifically, when they are related to a large variety of important aspects of singular behaviour of the solutions of such equations.
And such strange attractors are so often given by sets Γ with *positive* Lebesgue measure. Therefore, the usual theories of generalised functions cannot deal with them directly, that is, as singularity sets of their particular generalised functions. And this holds not only for Sobolev spaces or the Schwartz distributions, but also for the class of Colombeau type algebras, since the latter also require growth conditions on their particular generalised functions.

How to Introduce Large Classes of Singularities without Growth Conditions. Some of the above give a short description of the main facts which were faced, when the nowhere dense differential algebras of generalised functions were first considered. In particular, one would have liked to do away with each of the above three limitations.

The starting point in this regard has been there for a long time, namely, from the beginnings of the Schwartz distribution theory.
Indeed, a large enough nonlinear theory, containing the Schwartz distributions, could at least in principle be easily constructed, as soon as one was ready to deal with algebras, and not merely vector spaces of generalised functions. And the right, or natural, algebras to consider for that purpose where there, quite obviously. For instance, it is well known that the inclusion of the $\mathcal{C}^\infty$-smooth functions into the Schwartz distributions, namely

$$(12.1) \quad \mathcal{C}^\infty(\Omega) \subset \mathcal{D}'(\Omega)$$

is sequentially dense in the weak, and thus, strong topology. In other words, in a certain sense, we have the inclusion

$$(12.2) \quad \mathcal{D}'(\Omega) \;\text{"}\subset\text{"}\; (\mathcal{C}^\infty(\Omega)^{\mathbf{N}}$$

where the quotation marks mean that each element in the left hand term will correspond to many elements, more precisely, to many sequences of $\mathcal{C}^\infty$-smooth functions, in the right hand term.

In this way, it appears that the only problem left is to *remove* the quotation marks in (12.2). And this problem has well known precedents, the most basic being the construction of the real numbers $\mathbf{R}$ from the rational numbers $\mathbf{Q}$. Indeed, in this case again, one starts from the fact that $\mathbf{Q}$ is sequentially dense in $\mathbf{R}$, hence, in certain sense, one has the inclusion $\mathbf{R}$ " $\subset$ " $\mathbf{Q}^{\mathbf{N}}$. Here, the quotation marks are removed by the following *quotient space* construction

$$(12.3) \quad \mathbf{R} = \mathcal{A}/\mathcal{I}$$

where $\mathcal{A}$ is the subalgebra of Cauchy sequences in $\mathbf{Q}^{\mathbf{N}}$, while $\mathcal{I}$ is the ideal of $\mathcal{A}$ given by the sequences convergent to zero.

Now following the purely algebraic structure of the construction in (12.3), we can return to (12.2) and remove the quotation marks there in a similar manner. Namely, we choose

(12.4) $\quad \mathcal{I}(\Omega) \subset \mathcal{A}(\Omega) \subseteq (\mathcal{C}^\infty(\Omega))^{\mathbf{N}}$

where $\mathcal{A}(\Omega)$ is a subalgebra of $(\mathcal{C}^\infty(\Omega))^{\mathbf{N}}$ and $\mathcal{I}(\Omega)$ is an ideal in $\mathcal{A}(\Omega)$, such that (12.2) may become

(12.5) $\quad \mathcal{D}'(\Omega) \subset A = \mathcal{A}(\Omega)/\mathcal{I}(\Omega)$

In this case the quotient algebra A will be our algebra of generalised functions. And we note here that the introduction of a suitable differential structure on such an algebra will not present problems, see (7.1.8), (7.1.9).

Of course, we can try a *simpler* version of (12.4), (12.5), by choosing, see (7.1.3)

(12.6) $\quad \mathcal{A}(\Omega) = (\mathcal{C}^\infty(\Omega))^{\mathbf{N}}$

in which case $\mathcal{I}(\Omega)$ must be an ideal in $(\mathcal{C}^\infty(\Omega))^{\mathbf{N}}$. Clearly, if this construction may indeed work, this simpler version has the advantage that in (12.5), we now only have to find or construct *one single* entity, namely, the ideal $\mathcal{I}(\Omega)$.

And now we come to the main point related to the choice of the nowhere dense differential algebras of generalised functions.
Namely, in the simple case of (12.4) - (12.6), the ideal $\mathcal{I}(\Omega)$ will have to satisfy certain rather obvious *necessary* conditions. For instance, (12.1), (12.5), (12.6) immediately imply that we should like to have an *algebra embedding*

(12.7) $\quad \mathcal{C}^\infty(\Omega) \ni \psi \longmapsto T_\psi \in (\mathcal{C}^\infty(\Omega))^{\mathbf{N}}/\mathcal{I}(\Omega)$

and that the natural way for this embedding is to have

(12.8) $\quad T_\psi = (\psi, \psi, \psi, \ \ldots \ , \psi, \ \ldots \ ) + \mathcal{I}(\Omega), \quad \psi \in \mathcal{C}^\infty(\Omega)$

which is similar to what happens in (12.3), where we have the algebra, in fact, field embedding

(12.9) $\quad \mathbf{Q} \ni r \longmapsto (r, r, r, \ \ldots \ , r, \ \ldots \ ) + \mathcal{I} \in \mathcal{A}/\mathcal{I} = \mathbf{R}$

Now in view of (12.8), let us denote, see section 7.1

(12.10) $\quad \mathcal{U}(\Omega) = \{ \ (\psi, \psi, \psi, \ \ldots \ , \psi, \ \ldots \ ) \mid \psi \in \mathcal{C}^\infty(\Omega) \ \} \subset (\mathcal{C}^\infty(\Omega))^{\mathbf{N}}$

which is the set of sequences with each term the same $\mathcal{C}^\infty$-smooth function, and obviously, it is a subalgebra in $(\mathcal{C}^\infty(\Omega))^{\mathbf{N}}$, which is isomorphic with $\mathcal{C}^\infty(\Omega)$. Furthermore, it is the *diagonal* in the Cartesian product, more precisely, power $(\mathcal{C}^\infty(\Omega))^{\mathbf{N}}$. Now, the fact that (12.7) is an algebra embedding is equivalent with the following *off diagonality* condition, see (7.1.7)

(12.11) $\quad \mathcal{I}(\Omega) \cap \mathcal{U}(\Omega) = \{0\}$

So much for the strictly algebraic aspects of the construction of the differential algebras of generalised functions A.

And now, let us turn to the crucial issue of *singularities*, that is, to finding the *largest* possible class of singularity sets Γ in Ω, such that the differential algebras of generalised functions (12.4) - (12.6), (12.11) are able to deal with them.

First we note that the *lack* of flabbiness is a rather typical phenomenon which is not restricted to the Sobolev or Schwartz distributions, or to the Colombeau algerbras of generalised functions. Indeed, this lack of flabbiness happens as well with spaces of classical smooth functions. For instance, a $\mathcal{C}^\infty$-smooth function f_0 on a nonvoid open subset Ω_0 of Ω cannot in general be extended to a $\mathcal{C}^\infty$-smooth function f on Ω. And the same happens with the continuous functions as well.

Second, we can note that the *presence* of flabbiness can be closely related to the presence of a *large* class of singularities, namely, on all *closed, nowhere dense* subsets Γ in Ω. Indeed, as was shown in Oberguggenberger & Rosinger [pp. 142-146], in a certain natural sense, the smallest flabby sheaf which contains $\mathcal{C}^\infty(\Omega)$ is given by

$$(12.12) \quad \mathcal{C}^\infty_{nd}(\Omega) = \left\{ f : \Omega \to \mathbf{R} \;\middle|\; \begin{array}{c} \exists\, \Gamma \subset \Omega \text{ closed, nowhere dense } : \\ f \in \mathcal{C}^\infty(\Omega \setminus \Gamma) \end{array} \right\}$$

and a similar result holds for continuous functions as well.

In view of the above, we would like to construct the differential algebras of generalised functions A in (12.4) - (12.6), (12.11) - that is, we would like to choose the ideals $\mathcal{I}(\Omega)$ - in such a way that

$$(12.13) \quad \mathcal{C}^\infty_{nd}(\Omega) \subset A = (\mathcal{C}^\infty(\Omega))^{\mathbf{N}}/\mathcal{I}(\Omega)$$

Let us therefore try to see what is involved in (12.13). And we shall do so by analyzing the simplest possible nontrivial situation related to the mentioned kind of singularities, namely, when $\Gamma = \{0\} \subset \Omega = \mathbf{R}$, and we are given $f \in \mathcal{C}^\infty_{nd}(\Omega)$ which is defined by

$$(12.14) \quad f(x) = \left| \begin{array}{ll} 1 & \text{if } x < 0 \\ 1/x & \text{if } x > 0 \end{array} \right.$$

Clearly, because of its specific singularity at $x = 0$, the function f cannot be extended to a $\mathcal{C}^\infty$-smooth function on the whole of $\Omega = \mathbf{R}$. Nevertheless, through the desired embedding (12.13), we would like to preserve all the information on f which exists *outside* of its singularity, that is, on $\Omega \setminus \Gamma = \mathbf{R} \setminus \{0\}$.
Let us therefore note that, for every $a > 0$, there exist plenty of functions $f_a \in \mathcal{C}^\infty(\mathbf{R})$, such that, see (7.5.30) - (7.5.47)

$$(12.15) \quad f = f_a \quad \text{on} \quad \mathbf{R} \setminus (-a, a)$$

It follows that, when as an extension of (12.7), (12.8), we want to construct the embedding

$$(12.16) \quad f \longmapsto T_f = (\psi_0, \psi_1, \psi_2, \ldots, \psi_\nu, \ldots) + \mathcal{I}(\Omega) \in A = (\mathcal{C}^\infty(\Omega))^{\mathbf{N}} / \mathcal{I}(\Omega)$$

then in order to preserve the information on f outside its singularity, we can chose

$$(12.17) \quad \psi_\nu = f_{1/(\nu+1)}, \quad \nu \in \mathbf{N}$$

With this settled, we are now at the point to find out the appropriate structure of the desired ideals $\mathcal{I}(\Omega)$. Indeed, let us assume given any other equivalent representation of the embedding of f in (12.16), for instance

$$(12.18) \quad T_f = (\chi_\nu | \nu \in \mathbf{N}) + \mathcal{I}(\Omega)$$

then clearly

$$(12.19) \quad w = (\chi_\nu - \psi_\nu | \nu \in \mathbf{N}) \in \mathcal{I}(\Omega)$$

Therefore, in order not to disturb the information on f outside of its singularity, information which is already fully described in (12.16), (12.17), the terms in the sequence w above must eventually vanish in the neighbourhood of each point outside the singularity of f. And fortunately, this condition nearly manages to determine the structure of the desired ideals $\mathcal{I}(\Omega)$. Indeed, all we have to consider in addition is that the closed, nowhere dense singularity sets Γ should be allowed to be arbitrary in Ω.
In this way, we arrive at the following definition, see (7.1.1). The ideal $\mathcal{I}(\Omega)$ is the set of all sequences of functions

$$(12.20) \quad w = (w_\nu | \nu \in \mathbf{N}) \in (\mathcal{C}^\infty(\Omega))^{\mathbf{N}}$$

such that

$$(12.21) \quad \begin{array}{l} \exists\ \Gamma \subset \Omega \text{ closed, nowhere dense } : \\ \forall\ x \in \Omega \setminus \Gamma\ : \\ \exists\ \Delta \subseteq \Omega \setminus \Gamma \text{ neighbourhood of } x,\ \mu \in \mathbf{N}\ : \\ \forall\ \nu \in \mathbf{N},\ \nu \geq \mu\ : \\ \quad w_\nu = 0 \text{ on } \Delta \end{array}$$

We can note that *two* features distinguish the sequences of functions $w = (w_\nu | \nu \in \mathbf{N})$ in the ideal $\mathcal{I}(\Omega)$.
First, for each such sequence w, its terms w_ν, with $\nu \in \mathbf{N}$, have to *vanish* eventually on the neighbourhood of each point in an open, dense subset $\Omega \setminus \Gamma$, where Γ depends on the respective sequence w. This, as we have seen it, is the consequence of the fact that in the embeddings $f \longmapsto T_f$ in (12.13), (12.16), we want to *preserve* in T_f the whole information on f given on its set of nonsingular points.

Second, in view of the fact that the association $f \longmapsto f_a$ in (12.15) is multivalued, we have to allow a rather *arbitrary* behaviour for the respective functions f_a in the neighbourhood of the singularity of f. This implies that, given a sequence of functions $w = (w_\nu | \nu \in \mathbf{N})$ in the ideal $\mathcal{I}(\Omega)$, and its associated closed, nowehre dense subset Γ in Ω, we should *not* place any restrictions on the terms w_ν, with $\nu \in \mathbf{N}$, in the neighbourhood of Γ.

In short, the two clearly *opposite* distinguishing features of the sequences of functions w in the ideal $\mathcal{I}(\Omega)$ are that :

- outside their associated closed, nowhere dense subsets $\Gamma \subset \Omega$, they are *strongly controlled* by having to vanish eventually, while on the contrary

- in the neighbourhood of the respective Γ, they are *completely free*, as long as they are $\mathcal{C}^\infty$-smooth on the whole of Ω.

These ideals (12.20), (12.21), as we have seen in section 7.1., are called the nowhere dense ideals, and together with the nowhere dense differential algebras of generalised functions, see (1.6), (7.1.4), (7.1.3), (12.4) - (12.6), which they define, were first introduced and used in applications in Rosinger [3-5].
In the sequel, we return to the notation $\mathcal{I}_{nd}(\Omega)$, introduced in (7.1.1), for these nowhere dense ideals.

The Naturaleness of the Nowhere Dense Ideals. By now, there is only one question left about these nowhere dense ideals : except for the argument about the need to secure flabbiness, or more precisely, to obtain the embedding in (12.13), why should the subsets Γ of singularity be closed and nowhere dense ?

The answer to this question is as follows.
First, the singularity sets Γ should be closed, so that their complementaries $\Omega \setminus \Gamma$, that is, the sets of *nonsingularity*, can be open. Indeed, this goes along with many of the ways nonsingular points are defined. For instance, such is the situation when one defines the points at which functions are $\mathcal{C}^\infty$-smooth, or analytic. In other words, often it can become a rather difficult problem to define nonsingular points which do not have a neighbourhood of other nonsingular points.
Second, the singularity sets Γ have to be nowhere dense, since otherwise it is easy to see that the off diagonality condition (12.11) on the ideals $\mathcal{I}_{nd}(\Omega)$ will fail to be satisfied.

Flabbiness and the Largest Class of Singularities. It follows that, in the sense of (12.13), the nowhere dense ideals $\mathcal{I}_{nd}(\Omega)$, more precisely, the nowhere dense differential algebras of generalised functions A_Ω, which these ideals define through (12.4) - (12.6), (7.1.3), (7.1.4), are handling the *largest possible* class of singularities, namely, those given by arbitrary closed, nowhere dense subsets Γ, subsets which can have arbitrary large positive Lebesgue measure.
In this way, one obtains a framework for directly dealing with singularities such as those which may be described by large classes of fractal sets or strange attractors.

And as we have seen it above, these nowhere dense differential algebras of generalised functions can be determined through *two* rather simple and natural requirements.

First, we have those of a purely algebraic nature in (12.2) - (12.11), culminating in the last one, which is the so called off diagonality condition, and which is equivalent with the existence of the algebra embedding (12.7).
Second, we have the singularity condition (12.13) which is minimal in the sense that, as mentioned, $\mathcal{C}^\infty_{nd}(\Omega)$ is the smallest flabby sheaf containing $\mathcal{C}^\infty(\Omega)$, and thus allowing for singularities on arbitrary closed, nowhere dense subsets Γ in Ω.

As a further consequence, one can easily prove that the nowhere dense differential algebras of generalised functions form themselves a *flabby* sheaf, see Mallios & Rosinger.
This is unlike with many other spaces of generalised functions, see Kaneko, or with the Colombeau algebras.

How about the Distributions ? We should note that in (12.1), (12.2), the argument for the way to construct the nowhere dense differential algebras of generalised functions was strated with considerations related to the Schwartz distributions.
Soon after however, in (12.3) - (12.21), the Schwartz distributions did no longer appear.
Therefore, it is instructive to see in which way they can still be related to the mentioned algebras.

Let us recall here the algebraic characterisation in Rosinger [6] of very large classes of differential algebras of generalised functions A, for which the embeddings hold, see (12.5)

$$(12.22) \qquad \mathcal{D}'(\Omega) \subset A = (\mathcal{C}^\infty(\Omega))^{\mathbf{N}}/\mathcal{I}(\Omega)$$

where this time $\mathcal{I}(\Omega)$ denote arbitrary ideals in $(\mathcal{C}^\infty(\Omega))^{\mathbf{N}}$, and not only the nowhere dense ideal in (12.20), (12.21).
We also recall that these embeddings preserve the vector space structure of the Schwartz distributions, as well as the multiplication and differentiation of $\mathcal{C}^\infty$-smooth functions.

Quite surprisingly, the mentioned characterisation of the embeddings (12.22) comes down to the condition of off diagonality on the ideals $\mathcal{I}(\Omega)$ in (12.11), that is

$$(12.23) \qquad \mathcal{I}(\Omega) \cap \mathcal{U}(\Omega) = \{0\}$$

which is the same with the characterisation of the much *weaker* algebra embeddings (12.7), namely

$$(12.24) \qquad \mathcal{C}^\infty(\Omega) \subset A = (\mathcal{C}^\infty(\Omega))^{\mathbf{N}}/\mathcal{I}(\Omega)$$

The point here, however, is that the way the embeddings (12.22) are obtained in the characterisation in Rosinger [6], will be dependent on certain specific choices of *regularisations* of various usual representations of the Schwartz distributions.
And the only algebras of generalised functions known so far, where the embeddings (12.22) are made in a *natural* way, plus they preserve as well the differentiation of the Schwartz distributions, are the original and general Colombeau algebras, see Colombeau. In the later, simplifed versions of the Colombeau algebras, however,

versions which have been popular in applications, one again has to choose specific regularisations - through so called mollifiers - of the representation of Schwartz distributions.

In this way, it appears that the rather surprising equivalence between (12.22), (12.23) and (12.24), and the fact that, typically, the embeddings (12.22) are obtained through specific choices of regularisations of distributions, may be a reason for the mentioned lack of presence of the Schwartz distributions, noted in (12.3) - (12.21).

Furthermore, in the case of the nowhere dense algebras of generalised functions, we have already implemented the embedding (12.13), namely, in the form, see (7.1.4), (7.1.3)

(12.25) $\quad \mathcal{C}^{\infty}_{nd}(\Omega) = \mathcal{S}_{nd}(\Omega) \subset A_{\Omega}$

as seen in (7.5.30) - (7.5.48), and specifically, in (7.5.48).
However, in these embeddings again, the Schwartz distributions - more precisely, those of the Dirac and related types - tend to get out of the picture. Indeed, let us take once more a simplest nontrivial example, when $\Gamma = \{0\} \subset \Omega = \mathbf{R}$, and we consider the Heaviside function H, that is

$$H \in \mathcal{C}^{\infty}_{nd}(\Omega) = \mathcal{S}_{nd}(\Omega) \subset A_{\Omega}$$

Then clearly, see (7.5.30) - (7.5.47)

(12.26) $\quad H \neq 0 \in A_{\Omega}$

however, by applying to it the differentiation in the algebra A_{Ω}, we obtain, see (7.1.8), (7.1.9)

(12.27) $\quad \frac{d}{dx} H = 0 \in A_{\Omega}$

therefore, if with such a differentiation of generalised functions we still want to hold to the traditional relation

(12.28) $\quad \delta = \frac{d}{dx} H$

where δ denotes the Dirac distribution, then we shall obtain

(12.29) $\quad \delta = 0 \in A_{\Omega}$

By the way, this relation (12.29) does *not* mean that the nowhere dense differential algebras of generalised functions A_{Ω} will necessarily collapse the Dirac delta distributions into $0 \in A_{\Omega}$. Indeed, in the above particular, one dimensional case, for instance, we can take any $\psi \in \mathcal{C}^{\infty}(\Omega)$ such that

$$\psi(x) \neq 0, \quad x \in \Omega \qquad \text{and} \qquad \int_{\Omega} \psi(x) dx = 1$$

and for $\nu \in \mathbf{N}$, let us define $\psi_\nu \in \mathcal{C}^\infty(\Omega)$, by

$$\psi_\nu(x) = (\nu + 1)\psi((\nu + 1)x), \quad x \in \Omega$$

Then the sequence of functions $s = (\psi_\nu | \nu \in \mathbf{N})$ converges in $\mathcal{D}'(\Omega)$ to the Dirac δ distribution, that is

$$(12.30) \quad \lim_{\nu \to \infty} s = \lim_{\nu \to \infty} \psi_\nu = \delta \in \mathcal{D}'(\Omega)$$

and in view of (12.21), clearly $s \notin \mathcal{I}_{nd}(\Omega)$, therefore

$$(12.31) \quad S = s + \mathcal{I}_n(\Omega) \neq 0 \in A_\Omega$$

A similar situation, when any particular Schwartz distribution can be represented in many ways in a given algebra of generalised functions A in (12.5), is rather typical, and it is due to the mentioned need for the choice of particular regularisations of the distributions, in order to obtain these embeddings.

A Dichotomy. So far in the development of the algebraic nonlinear theories of generalised functions, we are facing a sort of dichotomy :

Either

I) We accept the embeddings (1.7), (12.5), (12.22) of the Schwartz distributions, embeddings which will typically depend on specific choices for the regularisation of distributions. This also includes the embeddings in (12.25), developed for the nowhere dense differential algebras of generalised functions A_Ω, algebras which, within this first alternative, have so far been the only ones used in a variety of applications, including those in the present work.
These embeddings (12.25), as seen above, enjoy a number of important advantages with respect to their capabilities in handling what appears to be the largest possible class of singularities. On the other hand, as seen in (12.27) - (12.29), the particular embeddings (12.25) of the Schwartz distributions into the algebras A_Ω leave out important relations and properties of the Schwartz distributions.

Or

II) We take the view that the embeddings of the Schwartz distributions into differential algebras of generalised functions should *maximally* preserve the properties of distributions. In particular, one is looking for a *natural* embedding of the Schwartz distributions into algebras of generalised functions. In this case, so far, we have to limit ourselves to the original and general Colombeau algebras, and therefore, accept the fact that those algebras are not a flabby sheaf, and cannot handle but the same traditional class of singularities, as the various linear theories of distributions have earlier done. Namely, they can only deal with singularities which must have zero Lebesque measure, and in the neighbourhood of which limitations are imposed by specific growth conditions.

The first alternative places the importance on the new, *nonlinear* aspects of generalised functions, and in the case of the nowhere dense differential algebras of generalised functions A_Ω, also on their ability to deal with what appears to be the *largest* possible class of singularities. At the same time, this first alternative is ready to accept that, so far, one could not find a transition between the linear distribution theory and the various nonlinear theories of generalised functions which would be free from all possible inconveniencies, and also, would allow the construction of algebras of generalised functions capable to handle the largest class of singularities.

The second alternative, on the other hand, appears to be strongly motivated by the *linear* tradition, as it places a high level of importance on the best possible connections between the linear theory of distributions and the nonlinear theories of generalised functions. In fact, this linear motivation is so strong that one does not quite ask the *question* whether, indeed, it is appropriate to condition, and thus limit, so severely the new nonlinear theories of generalised functions by the way they may connect to the earlier linear distribution theory.

Here it should be noted that, due to the relatively long time and well established tradition both in linear thinking in general, and in the use of the Sobolev and Schwartz linear theories of distributions in particular, there is at present a significant tendency towards taking the second alternative in the above dichotomy.
This popular choice is further motivated in the mind of its practitioners by the existence of a variety of rather natural structures of normed or locally convex topologies on the respective linear spaces of distributions, topologies which can be useful both in theoretical and numerical developments.

This motivation, however, has a *limited* validity. Indeed, it is often missed that *topological type phenomena* form a far larger world than that which can be described by the Kuratowski type *usual topologies*, or by their particular cases of metric spaces, unifrom spaces, normed spaces or locally convex spaces. For instance, in measure theory alone, fundamental results like the Lebesgue dominated convergence theorem, or the Egorov convergence theorem, cannot usefully be formulated in terms of usual topology. This is but a particular case of the fact that in ordered spaces, many relevant topological type phenomena, in particular, various concepts of order related convergence, cannot be expressed or characterised in a satisfatory form by using the usual concept of topology.
In fact, related to the study of linear and nonlinear PDEs, a certain tentative and limited awareness in this respect started to emerge, see Evans, and the respective literature cited there.
Strangely enough, this long time established stress on usual topologies has managed to fly in the face of particularly relevant, and also discomforting, contrary evidence, observed already in the early 1950s. Indeed, it was then noted that, in the case of general locally convex spaces, the usual bilinear functional which defines the duality, cannot be made bounded in the product topologies, see details in Kriegl & Michor.

On the other hand, the fact that, so far, usual topologies were not much used on the various algebras of generalised functions does not appear to handicap either the theoretical or numerical developments related to the solution of large classes

of linear and nonlinear PDEs, see for instance Biagioni, and the literature cited there.
Furthermore, it is *important* to realize that from *usual topological* point of view there is quite a *difference* between the various linear spaces of distributions, and on the other hand, the new algebras of generalised functions. Indeed, the former are typically defined as topological *duals* of spaces of classical and sufficiently smooth functions, and as such, they do *not* contain infinitesimal elements. The construction of the latter, however, is an obvious algebraic *generalisation* of the construction of the nonstandard reals $^*\mathbf{R}$. And in this way, in the case of the various algebras of generalised functions, one will again have *infinitesimals*, that is, generalised functions S, which in the sense of natural order relations, satisfy $0 < S < a$, for all $a \in (0, \infty)$.
The effect of the presence of these infinitesimal generalised functions is that usual Hausdorff topologies on the algebras of generalised functions will by necessity induce *discrete* topologies, when restricted to spaces of classical, sufficiently smooth functions. And clearly, this appears to be a rather unsatisfactory situation, since our everyday experience seems so strongly to suggest that, for such classical and sufficiently smooth functions, we do in fact have *nontrivial* neighbourhoods, made up from similar functions.
Now the conclusion from this sort of failure of usual topologies may merely be the obvious one, the one which appears to have been missed ever since the early 1950s. Namely that, on the algebras of generalised functions, instead of considering usual topologies, we should rather deal with thier natural and *more general* topological type phenomena. And one of such phenomena is already quite clear from the above :
That induced by various natural *order* structures which are inherent, like for instance, those which lead to the infinitesimal generalised functions.

However, regardless of all such facts, one need not stop enquiring deeper into the nature of the above dichotomy, nature which appears to bring us to rather novel and essential aspects of nonlinear mathematical modelling.
And it may well turn out that, as mentioned in chapter 1, the questioning by M. Hazewinkel of the need - or perhaps, merely comfort zone type habit - to hold so tight to, and to so much of the linear theory of distributions may in time acquire a new relevance, once we manage to progress deeper into the nature of the nonlinear phenomena involved in the study of generalised functions.

Nowhere Dense Algebras Applied in Abstract Differential Geometry and General Relativity. Finally, let us turn now to the second reason, mentioned at the beginning of this chapter, for the use in this work of the nowhere dense differential algebras of generalised functions A_Ω.

This is connected with some of their particularly successful recent applications, see Mallios & Rosinger, Mallios [6], based on the development of an abstract differential geometry in Mallios [1-5].
The main ideas underlying this abstract differential geometry are as follows.
Differential geometry, in particular, de Rham cohomology, is known to be one of the essential mathematical structures giving a deeper, intrinsic view of physics,

see Westenholz. Within mathematics itself, a major application of differential geometry, if not in fact, its very *'raisond'être'*, is in the study of PDEs and of their solutions. And it is precisely here that the usual differential geometry *collapses*, in view of the fact that nonlinear PDEs have both mathematically and physically relevant solutions which turn out to be *singular*.

Nevertheless, the usual way differential geometry is constructed relies heavily - and in fact, quite exclusively - on smooth functions. In particular, all the coefficients of differential forms are supposed to be smooth. And in this way, fundamental constructions, such as the de Rham complex, can only be made in the smooth case.

The effect is, of course, that usual differential geometry cannot deal with singularities.

In this way, as mentioned in the introduction, strating with the late 1970s, the idea emerged, Souriau [1,2], Heller [1,2], Kirillov [1,2], that, possibly, singularities should from the very *beginning* be incorporated into the spaces of functions - such as the coefficients of the differential forms, for instance - upon which the basic constructions of differential geometry are made. This program, however, owing to obvious reasons, could not be implemented by using the linear theory of distributions. Indeed, as mentioned earlier in this chapter, in addition to being merely linear and not nonlinear, the theory of distributions further exhibited a double limitation with respect to the singularities it could handle. First, such singularities had to be concentrated on subsets of zero Lebesgue measure, and second, the functions involved had to satisfy certain growth conditions in the beighbourhood of the singularities.

The abstract differential geometry introduced and developed in Mallios [1-5], starts from the recognition of the fact that, as it happened in other mathematical theories, such as for instance topology, the way of the historical emergence of differential geometry, ever since Gauss, may be such as to obscure the essential deeper structures involved. Indeed, the use of classical differential calculus, although had severely restricted the class of functions to which it could be applied, seemed for long to be quite clearly sine qua non in the construction of differential geometry. And yet, at a deeper analysis, it turns out that the truly important part is the *algebraic-operational* structure of the smooth functions in classical differential calculus. Furthermore, this algebraic-operational structure can to a large extent be recaptured in the far more general setting of certain suitable *sheaves of algebras of functions* on arbitrary *paracompact* spaces.

In this way, abstract differential geometry to a good extent becomes the *sheaf cohomology* of such sheaves of algebras.

The nowhere dense differential algebras of generalised functions A_Ω turn out to be particularly suitable such sheaves, in fact, flabby sheaves, see Mallios & Rosinger, upon which the construction of an abstract differential geometry can proceed. And with the use of these algebras A_Ω, all of the mentioned limitations of the linear theory of distributions fall away. In addition, this time, one can deal with singularities on arbitrary closed and nowhere dense sets, which therefore, can have arbitrary large positive Lebesgue measure.

In view of such recent physically directed applications of the nowhere dense differential algebras of generalised functions A_Ω, one has had a further reason to use them in this work, and develop on them arbitrary nonlinear, including nonprojectable Lie group actions.
And in fact, as seen in sections 7.2. - 7.4, 7.6. and 7.7., as well as in chapter 9, and specifically, in (7.3.1) - (7.3.8) and section 9.4, in order to define the mentioned kind of Lie group actions on them, these algebras themselves have to be further extended in ways which try to retain their topological type structures, rather than the algebraic ones. See also comments at the end of chapter 11.

Chapter 13

Lie Semigroup Actions and Semisymmetries

As mentioned at the beginning of chapter 1, and at the end of chapters 4 and 9, one of the rather unexpected significant advantages of the parametric approach to Lie group actions is the natural and easy way such an approach can be extended to Lie *semigroup* actions. And unlike studied so far in the literature, see Hilgert & Hofmann & Lawson, Hilgert & Neeb, Weinstein, this time, these Lie semigroup actions can be *genuine* semigroup actions, that is, they need *not* be subsemigroup actions of some Lie group actions. Also, one should note that the mentioned Lie semigroup actions in the literature cited are not applied to PDEs or their solutions.

Let us we further specify what we shall actually mean here by such genuine Lie semigroup actions $S \times M \longrightarrow M$ on open subsets $M \subseteq \mathbf{R}^l$. Namely, for the sake of simplicity, we shall ask that they satisfy the following two conditions :

A) They are one dimensional, and their action

(13.1) $\quad S \times M \longrightarrow M$

on M, see (1.1) - (1.5), can be identified with an action

(13.2) $\quad [0, \infty) \times M \longrightarrow M \quad \text{or} \quad [0, b) \times M \longrightarrow M$

where $([0, \infty), +)$ is the usual semigroup, while $([0, b), +)$ is the usual local semigroup, for $b > 0$.

B) The action (13.1) cannot be extended to a group action $\mathbf{R} \times M \longrightarrow M$, to a local group action $(-a, \infty) \times M \longrightarrow M$, or to a local group action $(-a, b) \times M \longrightarrow M$, where $(\mathbf{R}, +)$ is the usual group, while for $a, b > 0$, $((-a, \infty), +)$ and $((-a, b), +)$ are the usual local groups, Olver [1,2].

In other words, these two conditions A) and B) mean that we shall only consider Lie semigroup actions (13.1) on M which staisfy the follwing condition of being *genuine Lie semigroups* :

(GLS). The Lie semigroup actions are one dimensional, and no other action in the semigroups, except for the identical action alone, has an inverse.

As also mentioned earlier, the fact that the way we defined in chapter 4 parametric Lie group actions can be used without any modification, in order to define genuine Lie semigroup actions as well, follows in a most simple and direct manner from the functorial advantage of the parametric approach, and its root is in the natural definition of parametric Lie group actions in (4.1) - (4.4). Indeed, given the Lie group setup in (1.1) - (1.5), it is rather trivial to note that, in order to be able to define (4.2) through (4.3), the mapping g in that latter commutative diagram need *not* be a $\mathcal{C}^\infty$-smooth diffeomorphism of M, as it suffices that $g \in \mathcal{C}^\infty(M, M)$.

To formulate the issue more systemtically, we note that (1.5) means the existence of a group homomorphism

$$(13.3) \quad G \ni g \longrightarrow \mathcal{D}iff^\infty(M)$$

where $\mathcal{D}iff^\infty(M)$ denotes the noncommutative group, under the usual composition of functions, of the $\mathcal{C}^\infty$-smooth diffeomorphisms of M. Clearly, for nontrivial M, we have the *strict* inclusion

$$(13.4) \quad \mathcal{D}iff^\infty(M) \subset \mathcal{C}^\infty(M, M)$$

and as is known, the right hand term is significantly larger than the left hand term. This is clear even in the particular case of $M = \mathbf{R}$, while in the general case, one can see related facts in Michor, Kriegl & Michor. Moreover, under the usual composition of functions, the right hand term in (13.4) is a noncommutative semigroup with identity, and as such, it cannot be the subsemigroup of any group, since it contains noninvertible functions.
In this way, it follows that the *parametric* action (4.1) on functions in $\mathcal{C}^\infty_n(M)$ can be defined not only for g in the group $\mathcal{D}iff^\infty(M)$, but also for g in the much larger and genuine semigroup $\mathcal{C}^\infty(M, M)$.

The aim of this chapter is to present a first introduction, as well as a motivation, of a theory of genuine Lie semigroup actions on open sets M, actions which, in terms of (13.4), belong to the semigroup in the right hand term of that relation, and cannot belong to the group in its left hand term.

In fact, owing to its short and heuristic-exploratory nature, what is presented here will rather be a mere 'introduction to a future introduction' to the mentioned theory of genuine Lie semigroup actions.

From the start, let us note here that a significant interest in this kind of extension to genuine Lie semigroup actions can come from the following two facts.
First, as already mentioned, the class of such Lie semigroup actions is far *larger* than that of Lie group actions. And as we have seen, one simple way to suggest the extent of such a difference between these two classes of actions is to compare it to the difference between the group of $\mathcal{C}^\infty$-smooth diffeomorphisms of M, and on the other hand, the much larger semigroup of $\mathcal{C}^\infty$-smooth mappings from M to M.

Second, following from the above, one can obtain a far more *powerful* tool in the study of solutions of nonlinear PDEs. Indeed, now one can apply to such solutions the far larger class of Lie semigroup actions, and find their invariance properties in this extended framework of what we shall call *semisymmetries*. And in order to illustrate such semisymmetries, we shall start this chapter by giving a few rather simple examples.

In section 8, as suggested in a private correspondence by P J Olver, we shall indicate a few other situations of interest where Lie semigroups appear in a natural manner.

1. Examples of Semisymmetries of Solutions of Nonlinear PDEs

The aim of this section is merely to illustrate the existence of actions on the solutions of certain PDEs, actions which turn solutions into solutions, and which nevertheless are generated by $\mathcal{C}^\infty$-smooth mappings which *fail* to be injective, thus cannot be $\mathcal{C}^\infty$-smooth diffeomorphisms.

Let us therefore start with considering the simple linear PDE

$$(13.1.1) \quad U_t = U_x, \quad (t,x) \in \Omega = \mathbf{R}^2$$

in which case, with the notation in (1.1) - (1.5), we have $M = \Omega \times \mathbf{R} = \mathbf{R}^3$. The $\mathcal{C}^\infty$-smooth solutions of (13.1.1) are given by

$$(13.1.2) \quad U(t,x) = h(t+x), \quad (t,x) \in \Omega$$

where $h \in \mathcal{C}^\infty(\mathbf{R},\mathbf{R})$ is arbitrary.
Let us take now any $f \in \mathcal{C}^\infty(M,M)$ which is of the form

$$(13.1.3) \quad (t,x,u) \overset{f}{\longmapsto} (t,x,a(u))$$

with $a \in \mathcal{C}^\infty(\mathbf{R},\mathbf{R})$ an arbitrary given function which is not injective.
Clearly, the resulting f in (13.1.3) will fail to be injective, and thus a $\mathcal{C}^\infty$-smooth diffeomorphism of M.
And yet, if we act on the solution U in (13.1.2) by f in (13.1.3), then we shall obtain

$$(13.1.4) \quad \bar{U}(t,x) = (f\ U)(t,x) = a(h(t+x)), \quad (t,x) \in \Omega$$

and obviously $\bar{U} = f\ U$ is again a solution of (13.1.1).

As another simple example, let us consider any nonlinear PDE in (1.1), (1.4), for which the constant functions $U(x) = c,\ x \in \Omega,\ c \in \mathbf{R}$ are solutions, that is

$$(13.1.5) \quad T(x,D)\ c = 0 \ \text{ on } \Omega$$

We take now $f : M \to M$ given by

$$(13.1.6) \quad (x,u) \overset{f}{\longmapsto} (x, a(u))$$

where $a \in \mathcal{C}^\infty(\mathbf{R}, \mathbf{R})$ is an arbitrary function which is not injective. Then again $f \in \mathcal{C}^\infty(M, M)$ will not be injective, and thus it will fail to be a $\mathcal{C}^\infty$-smooth diffeomorphism of M.
If now we act on the constant solution $U = c$ of (13.1.5) by f in (13.1.6), then we obtain

$$(13.1.7) \quad \bar{U}(x) = (f\ U)(x) = a(c), \quad x \in \Omega$$

which is another constant solution of (13.1.5).

The above examples of semisymmetries of solutions are no doubt quite simple, however, they clearly illustrate the main issue here. Namely, the two actions $f\ U$ in (13.1.4) and (13.1.7) cannot be Lie group actions, since they are defined respectively by the actions (13.1.3) and (13.1.6), which are not invertible, hence, they cannot be $\mathcal{C}^\infty$-smooth diffeomorphisms.
And yet, it is possible to act with them on solutions. And when one does so, one again obtains solutions.
On the other hand, the actions (13.1.3) and (13.1.6) are $\mathcal{C}^\infty$-smooth mappings of the respective M into themselves. Thus each of them belongs to the semigroup $\mathcal{C}^\infty(M, M)$ of all such mappings. It follows that, if we take any such action f, then together with id_M, it will generate a subsemigroup

$$(13.1.8) \quad S_f \subset \mathcal{C}^\infty(M, M)$$

Therefore, a possible formulation of the problem is to find semigroups S_f in such a way that they can be extended to one dimensional semigroups which will satisfy condition (GLS).
This problem will be addressed, starting with the next section.

We note that the above argument on actions on M which belong to the semigroup $\mathcal{C}^\infty(M, M)$ but not to the group $\mathcal{D}iff^\infty(M)$, justifies the name of semisymmetries, when they act on solutions of ODEs or PDEs.

Let us now present a few somewhat more involved examples, before in the next sections, we shall try to go into the beginnings of the theory.

Let us consider the nonlinear PDE

$$(13.1.9) \quad UU_{tx} - U_t U_x = 0, \quad (t,x) \in \Omega = \mathbf{R}^2$$

which has solutions given by

$$(13.1.10) \quad U(t,x) = h(t)k(x), \quad (t,x) \in \Omega$$

for any $h,\ k \in \mathcal{C}^\infty(\mathbf{R}, \mathbf{R})$.
We define the $\mathcal{C}^\infty$-smooth action $f : M \to M$, where $M = \Omega \times \mathbf{R} = \mathbf{R}^3$, by

$$(13.1.11) \quad (t,x,u) \overset{f}{\longmapsto} (t,x,u^l)$$

for any given $l \in \mathbf{N}$, $l \geq 2$, l even. Clearly, f is then never injective, hence it always fails to be a $\mathcal{C}^\infty$-smooth diffeomorphism.
However, by acting with f on the solutions (13.1.10), we obtain

$$(13.1.12) \quad \bar{U}(t,x) = (f\ U)(t,x) = h^l(t)k^l(x), \quad (t,x) \in \Omega$$

which are also solutions of (13.1.9).

One more example is the following. Let the nonlinear PDE

$$(13.1.13) \quad F(U_t, U_x) = 0, \quad (t,x) \in \Omega = \mathbf{R}^2$$

be defined by any given function $F \in \mathcal{C}^\infty(\mathbf{R}^2, \mathbf{R})$, which is asuumed to be homogeneous of a certain fixed degree.
We shall consider actions f on M as in (13.1.3), which as mentioned, are not invertible, hence cannot be $\mathcal{C}^\infty$-smooth diffeomorphisms of M.
However, due to the homogeneity of F, whenever $U \in \mathcal{C}^\infty(\Omega, \mathbf{R})$ is a solution of (13.1.13), so will be the result $f\ U$ of the action of f on U, namely

$$(13.1.14) \quad \bar{U}(t,x) = (f\ U)(t,x) = a(U(t,x)), \quad (t,x) \in \Omega$$

Let us finally note here that in rather simple ways, one can endlessly complicate the above list of examples of PDES which have semisymmetries of their solutions. Indeed, let $T(x,D)$, $S(x,D)$ and $Q(x,D$ be arbitrary nonlinead operators such as in (1.4). It is then obvious that, if $U \in \mathcal{C}^\infty(\Omega, \mathbf{R})$ is a solution of the nonlinear PDE

$$(13.1.15) \quad T(x,D)\ U(x) = S(x,D)\ U(x), \quad x \in \Omega$$

then it is also a solution of the more complicated and higher order nonlinear PDE

$$(13.1.16) \quad Q(x,D)T(x,D)\ U(x) = Q(x,D)S(x,D)\ U(x), \quad x \in \Omega$$

And clearly, this observation allows the construction of higher order and more complicated examples of PDEs which have nontrivial semisymmetries of solutions, by simply applying to the earlier PDEs, such as for instance those above, any partial differential operators $Q(x,D)$.

2. How to Generate Genuine Lie Semigroup Actions

We shall now go beyond the rather simple examples of semisymmetries in the previous section, and explore a method for constructing genuine Lie semigroup actions which satisfy the condition (GLS).
This method is an immediate extension of the classical way of constructing one dimensional Lie group actions. Such Lie group actions are obtained as $\mathcal{C}^\infty$-smooth flows on given open subsets $M \subseteq \mathbf{R}^l$, and are defined by autonomous ODEs, which are determined by the infinitesimal generators $F \in \mathcal{C}^\infty(M, \mathbf{R}^l)$ of the respective

Lie group actions. For convenience, let us shortly recall here the essential features of this classical construction.

Lie Groups, Actions, Infinitesimal Generators, ODEs, Flows and Evolution Operators. Given any mapping $F \in \mathcal{C}^\infty(M, \mathbf{R}^l)$, we can associate with it the autonomous nonlinear system of ODEs

$$(13.2.1) \quad \begin{aligned} &\frac{d}{d\epsilon} Y(\epsilon) = F(Y(\epsilon)), \quad \epsilon \in \mathbf{R} \\ &Y(0) = y \in M \end{aligned}$$

Then the unique solution, assumed here for simplicity to exists globally

$$(13.2.2) \quad \mathbf{R} \ni \epsilon \longrightarrow Y(\epsilon) \in M$$

defines a flow on M, through each point $y \in M$, a flow which corresponds to a one dimensional Lie group G which acts on M according to

$$(13.2.3) \quad M \ni y \stackrel{\epsilon}{\longmapsto} Y(\epsilon) = y' \in M$$

In this case, F is called the infinitesimal generator of the Lie group action (13.2.3).

An essential fact in the classical Lie theory is that the converse of the above construction also holds. Namely, if we are given a one dimensional Lie group G which acts on M according to (13.2.3), then we can define its infinitesimal generator $F \in \mathcal{C}^\infty(M, \mathbf{R}^l)$, as the mapping

$$(13.2.4) \quad F(y) = \frac{d}{d\epsilon} Y(\epsilon)|_{\epsilon=0}, \quad y \in M$$

and repeating the steps (13.2.1) - (13.2.3), we shall get back the one dimensional Lie group G, which acts on M, according to (13.2.3), that is, the action with which we started.

It is convenient here to introduce the concept of evolution operator E associated with (13.2.1) - (13.2.4), namely

$$(13.2.5) \quad \begin{aligned} &\mathbf{R} \ni \epsilon \longmapsto E(\epsilon) : M \longrightarrow M \\ &E(\epsilon)(y) = Y(\epsilon), \quad \epsilon \in \mathbf{R}, \ y \in M \end{aligned}$$

Clearly, the above one dimensional Lie group action (13.2.3) can now be identified with

$$(13.2.6) \quad \mathbf{R} \times M \ni (\epsilon, y) \longmapsto E(\epsilon)(y) = Y(\epsilon) \in M$$

Finding Infinitesimal Generators for Genuine Lie Semigroups. In view of (13.2.1) - (13.2.6), we shall start our attempts at finding genuine Lie semigroup actions with the simple one dimensional Lie semigroup actions, given by

$$(13.2.7) \quad [0, \infty) \times M \longrightarrow M$$

The problem, therefore, is how to find the respective infinitesimal generators which, through the solutions of the associated ODEs, would give genuine Lie semigroup actions (13.2.7).
For that purpose, here we shall pursue a somewhat heuristic approach which, step by step, may both motivate and clarify the method, as well as its results.

One natural way to start is to try to extend the above construction (13.2.1) - (13.2.6) in such a manner that we avoid obtaining group or local group actions. This means that, among others, we avoid obtaining invertible actions, except for the identical one.
Of course, this first of all means that our genuine Lie semigroup infinitesimal generators F have to satisfy

$$(13.2.8) \quad F \notin \mathcal{C}^\infty(M, \mathbf{R}^l)$$

since otherwise the situation in (13.2.1) - (13.2.6) would inevitably follow, and we would be back to Lie group or local group actions.
Now, a simple way to secure condition (13.2.8) is to include certain kind of *singularities* in F, namely, to ask that

$$(13.2.9) \quad F : (M \setminus \Sigma) \to \mathbf{R}^l, \quad F \in \mathcal{C}^\infty(M \setminus \Sigma, \mathbf{R}^l)$$

for suitable $\Sigma \subset M$ which is the singularity set of F. However, as seen next in a simple example, this need not always work. Indeed, let us take $M = \mathbf{R}$ and consider the ODE for the function $Y : \mathbf{R} \to M$, given by

$$\frac{d}{d\epsilon} Y(\epsilon) = 1/(Y(\epsilon))^2, \quad \epsilon \in \mathbf{R}$$

$$Y(0) = y \in M \setminus \{0\}$$

Here of course $F(y) = 1/(y)^2, \quad y \in M \setminus \Sigma$, with the notation in (13.2.9), where $\Sigma = \{0\} \subset M$. However, the unique $\mathcal{C}^\infty$-smooth solution of this ODE, for $y \in M \setminus \Sigma$, is

$$Y(\epsilon) = \sqrt[3]{3\epsilon + y^3} \in M, \quad \epsilon \in \mathbf{R}, \quad \epsilon \neq -y^3/3$$

Thus it follows that our F, although it satisfies (13.2.8), (13.2.9) with a nonvoid singularity set Σ, nevertheless defines a local Lie group action. Therefore, there will be actions other than the identity, and which will have an inverse.
We conclude that autonomous ODEs, even if they are singular, need not define genuine Lie semigroup actions.

The next step we can try is to replace the autonomous $\mathcal{C}^\infty$-smooth ODEs in (13.2.1), with nonautonomous $\mathcal{C}^\infty$-smooth ODEs.
Here however, we face the fact that, as shown in the next section, nonautonomous systems of ODEs can easily be reduced to autonomous ones, simply by adding one more rather trivial ODE to them. This means that the *nonautonomous* ODEs we

have to consider must in a certain suitable sense be *singular* at $\epsilon = 0$, together with their solutions.

A Simple Example of Singular Nonautonomous ODE and its Solutions. Let us consider a simple example in this regard. We take again $M = \mathbf{R}$, while the function $H \in \mathcal{C}^0([0,\infty) \times M, M) \cap \mathcal{C}^\infty((0,\infty) \times M, M)$ will be defined by

$$(13.2.10) \quad H(\epsilon, y) = y + \sqrt{\epsilon}\, y^2, \quad \epsilon \in [0,\infty),\ y \in M$$

Thus we have

$$H(0,y) = y, \quad y \in M$$

$$(13.2.11) \quad H(\epsilon,\ .\ ) \notin \mathcal{D}iff^\infty(M), \quad \epsilon \in (0,\infty)$$

$$H \notin \mathcal{C}^1([0,\infty) \times M, M)$$

Clearly, H defines the following action on M

$$(13.2.12) \quad [0,\infty) \times M \ni (\epsilon, y) \longmapsto H(\epsilon, y) \in M$$

and this cannot be part of a group, or local group action on M, since as we have seen in (13.2.11), the mapping $H(\epsilon,\ .\ )$, with $\epsilon \in (0,\infty)$, is not a $\mathcal{C}^\infty$-smooth diffeomorphism of M.

Let us, therefore, see what, if any, is the ODE which is satisfied by H, when taken as a function of ϵ. In other words, let us find what would correspond as, see (13.3.7) below, the nonautonomous infinitesimal generator F of the action on M defined in (13.2.12) by H.
We note that (13.2.10) gives

$$(13.2.13) \quad \frac{\partial}{\partial \epsilon} H(\epsilon, y) = \frac{y^2}{2\sqrt{\epsilon}}, \quad \epsilon \in (0,\infty),\ y \in M$$

Now we simply solve (13.2.10) as a quadratic equation in y, and obtain two solutions

$$(13.2.14) \quad y_{1,2} = \frac{-1 \pm \sqrt{1 + 4\sqrt{\epsilon} H}}{2\sqrt{\epsilon}}, \quad \epsilon \in (0,\infty),\ 1 + 4\sqrt{\epsilon} H \geq 0$$

Further, we return to (13.2.10), replace y with its values from (13.2.14), while y^2 will be replaced with its value computed from (13.2.13). Thus we obtain two nonautonomous singular ODEs

$$(13.2.15) \quad \frac{\partial}{\partial \epsilon} H = \frac{1 + 2\sqrt{\epsilon} H \pm \sqrt{1 + 4\sqrt{\epsilon} H}}{4\epsilon\sqrt{\epsilon}}, \quad \epsilon \in (0,\infty),\ 1 + 4\sqrt{\epsilon} H \geq 0$$

The first one of them, corresponding to the sign ' - ' in the front of the radical, can be written as

$$\frac{\partial}{\partial \epsilon} Y(\epsilon) = \frac{Y(\epsilon)^2}{\sqrt{\epsilon}\,(1 + 2\sqrt{\epsilon}Y(\epsilon) + \sqrt{1 + 4\sqrt{\epsilon}Y(\epsilon)}\,)} \tag{13.2.16}$$

$$\epsilon \in (0, \infty),\ 1 + 4\sqrt{\epsilon}Y(\epsilon) \geq 0$$

while the second one is

$$\frac{\partial}{\partial \epsilon} Y(\epsilon) = \frac{1 + 2\sqrt{\epsilon}Y(\epsilon) + \sqrt{1 + 4\sqrt{\epsilon}Y(\epsilon)}}{4\epsilon\sqrt{\epsilon}} \tag{13.2.17}$$

$$\epsilon \in (0, \infty),\ 1 + 4\sqrt{\epsilon}Y(\epsilon) \geq 0$$

Here we note that in (13.2.15), we have considered H given in (13.2.10) as being a function of $\epsilon \in (0, \infty)$, while $y \in M$ is arbitrary, but fixed. With this meaning given to H, now it is easy to see that, in case $(\epsilon, y) \in (0, \infty) \times M$ is such that

$$1 + 2\sqrt{\epsilon}y \geq 0 \tag{13.2.18}$$

then the ODE in (13.2.16) is satisfied by H, on the other hand, when

$$1 + 2\sqrt{\epsilon}y \leq 0 \tag{13.2.19}$$

then H will satisfy the ODE in (13.2.17). Moreover, both ODEs, namely (13.2.18), (13.2.16), as well as (13.2.19), (13.2.17) will for H specified as above be associated with the initial condition

$$\lim_{\epsilon \to 0} Y(\epsilon) = y \in M \tag{13.2.20}$$

where we note that the ODEs (13.2.16), (13.2.17) being only defined for $\epsilon > 0$, we cannot ask for an initial condition of the form $Y(0) = y \in M$.

This example suggests that, in the context of (1.1) - (1.5), when we are looking for genuine Lie semigroup actions, we should take as infinitesimal generators certain functions $F \in \mathcal{C}^\infty((0, \infty) \times M, \mathbf{R}^l)$, and then associate with each of them the corresponding *nonautonomous* system of nonlinear ODEs

$$\begin{aligned} &\frac{d}{d\epsilon} Y(\epsilon) = F(\epsilon, Y(\epsilon)), \quad \epsilon \in (0, \infty) \\ &\lim_{\epsilon \to 0} Y(\epsilon) = y \in M_0 \end{aligned} \tag{13.2.21}$$

for a suitable subset $M_0 \subseteq M$, system which together with its relevant solutions should in a certain sense be *singular* at $\epsilon = 0$. Clearly, the above initial condition on Y, at $\epsilon = 0$, is the correspondent form of the initial condition in (13.2.20). Further, we should assume that the unique solutions of such ODEs satisfy

$$\begin{aligned} &*)\ \ Y \in \mathcal{C}^\infty((0, \infty), M) \\ &**)\ \ Y \in \mathcal{C}^0([0, \infty), M) \setminus \mathcal{C}^1([0, \infty), M) \end{aligned} \tag{13.2.22}$$

and also

(13.2.23) $(0, \infty)$ is maximal in $\mathbf{R}$ for $*)$ in (13.2.22)

A Class of more General Nonautonomous Singular ODE Systems and Solutions. We end this section by giving a more general formulation of the construction in the above example (13.2.10) - (13.2.20). The obvious interest in this class comes from the fact that its various cases can incorporate *every* semigroup action from $\mathcal{C}^\infty(M, M) \setminus \mathcal{D}iff^\infty(M)$, see (13.2.28) below.

Let us take for simplicity $M = \mathbf{R}^l$. Further, let us take any

(13.2.24) $f \in \mathcal{C}^\infty(M, M) \setminus \mathcal{D}iff^\infty(M)$

We also take any auxilliary function $\alpha \in \mathcal{C}^0([0, \infty), \mathbf{R}) \cap \mathcal{C}^\infty((0, \infty), \mathbf{R})$ such that

$$\alpha(0) = 0, \quad \alpha(1) = 1$$

$$(13.2.25) \qquad \alpha \notin \mathcal{C}^1([0, \infty), \mathbf{R})$$

$$\alpha'(\epsilon) \neq 0, \quad \epsilon \in (0, \infty)$$

Now we can define

(13.2.26) $H \in \mathcal{C}^0([0, \infty) \times M, M) \cap \mathcal{C}^\infty((0, \infty) \times M, M)$

as the homotopical deformation of id_M into f, namely

(13.2.27) $H(\epsilon, y) = (1 - \alpha(\epsilon)\,)y + \alpha(\epsilon) f(y), \quad \epsilon \in [0, \infty), \; y \in M$

Then obviously

(13.2.28) $H(0, \,.\,) = id_M, \quad H(1, \,.\,) = f(\,.\,)$

hence for certain values of ϵ, like for instance, $\epsilon = 1$, we shall have that

(13.2.29) $H(\epsilon, \,.\,) \notin \mathcal{D}iff^\infty(M)$

Sometime, we may as well assume the stronger property, see (13.2.11)

(13.2.30) $H(\epsilon, \,.\,) \notin \mathcal{D}iff^\infty(M), \quad \epsilon \in (0, \infty)$

However, as seen in the example at the end of section 4, condition (13.2.30) is *not* necessary, in order to obtain a genuine Lie semigroup action in (13.2.31) next. Clearly, H in (13.2.27) defines an action, see (13.2.12)

(13.2.31) $[0, \infty) \times M \ni (\epsilon, y) \longmapsto H(\epsilon, y) \in M$

Now, as in the example (13.2.10) - (13.2.20), we shall try to find a nonautonomous system of ODEs which H satisfies as a function of ϵ. That is, we want to find the infinitesimal generator F for a system which may be like that in (13.2.21) - (13.2.23).

For that purpose, we shall proceed in a way similar to that in the mentioned example, and from (13.2.27) we deduce the system

(13.2.32) $H = (1-\alpha(\epsilon))y+\alpha(\epsilon)f(y), \quad \frac{\partial}{\partial\epsilon}H = \alpha'(\epsilon)(f(y)-y), \quad \epsilon \in \mathbf{R},\ y \in M$

which are linear algebraic equation in y and $f(y)$. Since $\alpha'(\epsilon) \neq 0$ according to (13.2.25), we can solve (13.2.32) for y and $f(y)$. In this way, we obtain

$$y = \frac{\alpha' H - \alpha H_\epsilon}{\alpha'}, \quad f(y) = \frac{(1-\alpha)H_\epsilon + \alpha' H}{\alpha'}$$

and therefore, also the implicit nonautonomous system of ODEs which is satisfied by H as a function of ϵ, namely

(13.2.33)
$$(1-\alpha(\epsilon))\frac{d}{d\epsilon}Y(\epsilon) + \alpha'(\epsilon)Y(\epsilon) = \alpha'(\epsilon) f\left(\frac{\alpha'(\epsilon)Y(\epsilon) - \alpha(\epsilon)\frac{d}{d\epsilon}Y(\epsilon)}{\alpha'(\epsilon)}\right),$$
$$\epsilon \in (0,\infty)$$

which is singular at $\epsilon = 0$. Thus in view of (13.2.28), we can associate with it the initial condition, see (13.2.20)

(13.2.34) $\lim_{\epsilon\to 0} Y(\epsilon) = y \in M$

Clearly, far more involved general examples than that in (13.2.24) - (13.2.34) can be considered. For instance, in case we want to avoid the restriction $\alpha'(\epsilon) \neq 0,\ \epsilon \in (0,\infty)$ in (13.2.25) on the derivative of α, we can proceed as follows. We start with f in (13.2.24). Then we take any α as in (13.2.25), which instead of the above restriction on its derivative, satisfies the condition

(13.2.35) $\alpha(\epsilon) \neq 0, \quad \epsilon \in (0,\infty)$

Further, we take any $\beta \in \mathcal{C}^\infty([0,\infty),\mathbf{R})$. Then instead of using (13.2.27), we can define H in (13.2.26) by

(13.2.36) $H(\epsilon, y) = (1-\alpha(\epsilon))y + \alpha(\epsilon)f(\beta(\epsilon)y), \quad \epsilon \in [0,\infty),\ y \in M$

Let us give a simple example of this general case. We take again $M = \mathbf{R}$, and correspondingly, f as in (13.2.24). Let α be given by

(13.2.37)
$$\alpha(\epsilon) = c\epsilon(2+\sin(\tfrac{1}{\epsilon})), \quad \epsilon \in (0,\infty)$$
$$\alpha(0) = 0, \quad \alpha(1) = 1$$

while $\beta \in \mathcal{C}^\infty([0,\infty),\mathbf{R})$ is such that

(13.2.38) $\beta(1) = 1, \quad \text{and} \quad \beta(\epsilon),\ \beta'(\epsilon) \neq 0, \quad \epsilon \in [0,\infty)$

Then

(13.2.39) $\frac{\partial}{\partial\epsilon}H(\epsilon,y) = -\alpha'(\epsilon)y + \alpha'(\epsilon)f(\beta(\epsilon)y) + \alpha(\epsilon)\beta'(\epsilon)yf'(\beta(\epsilon)y), \quad \epsilon \in (0,\infty)$

and it is easy to see that, in general, that is, for a large class of f and β, one can eliminate y from (13.2.36), (13.2.39), and similar to (13.2.33), one obtains an implicit nonautonomous ODE, singular at $\epsilon = 0$, and which is satisfied by H as a function of ϵ and with $y \in M$ arbitrary fixed, namely

(13.2.40) $\quad W(\epsilon, Y(\epsilon), \frac{\partial}{\partial \epsilon} Y(\epsilon)) = 0, \quad \epsilon \in (0, \infty)$

And again, this system of ODEs will be associated with the intital condition, see (13.2.34)

(13.2.41) $\quad \lim_{\epsilon \to 0} Y(\epsilon) = y \in M$

Nonremovable Singularities. It is important to note here that the property

(13.2.42) $\quad H \in \mathcal{C}^0([0, \infty) \times M, M) \setminus \mathcal{C}^1([0, \infty) \times M, M)$

in (13.2.10), (13.2.11), (13.2.26), (13.2.27), (13.2.36) - (13.2.38) is needed, in order that the respective ODEs in (13.2.16), (13.2.17), and more generallly, in (13.2.21), (13.2.40), may have solutions with *nonremovable* singularities at $\epsilon = 0$, that is, solutions which satisfy (13.2.23). And as we shall see in section 4, this is precisely why such ODEs, more exactly, their autonomous associates, will through their solutions define *genuine* Lie semigroup actions.

Let us make one more remark related to this issue of nonremovable singularities of ODEs and their solutions.
Obviously, with H in (13.2.10), we can associate $K \in \mathcal{C}^\infty(\mathbf{R} \times M, M)$ defined by

(13.2.43) $\quad K(\omega, y) = y + \omega y^2, \quad \omega \in \mathbf{R}, \quad y \in M$

in the sense that

(13.2.44) $\quad H(\omega^2, \ . \ ) = K(|\omega|, \ . \ ), \quad H(\epsilon, \ . \ ) = K(\sqrt{\epsilon}, \ . \ ), \quad \omega \in \mathbf{R}, \quad \epsilon \in [0, \infty)$

Furthermore, $K(\ . \ , y)$, for any fixed $y \in M$, will satisfy the nonautonomous singular ODEs

(13.2.45)
$$\frac{d}{d\omega} Z(\omega) = \frac{2Z^2(\omega)}{1 + 2\omega Z(\omega) + \sqrt{1 + 4\omega Z(\omega)}},$$
$$\omega \in \mathbf{R}, \quad 1 + 4\omega Z(\omega) \geq 0$$

respectively

(13.2.46)
$$\frac{d}{d\omega} Z(\omega) = \frac{1 + 2\omega Z(\omega) + \sqrt{1 + 4\omega Z(\omega)}}{2\omega^2}$$
$$\omega \in \mathbf{R}, \quad 1 + 4\omega Z(\omega) \geq 0$$

accordign to whether $(\omega, y) \in \mathbf{R} \times M$ satisfies the condition

(13.2.47) $\quad 1 + 2\omega y \geq 0$

or the condition

$$(13.2.48) \qquad 1 + 2\omega y \leq 0$$

And in both cases, K will satisfy the intial condition

$$(13.2.49) \qquad K(0, y) = y \in M$$

However, although the ODE in (13.2.46) is singular at $\omega = 0$, its solution for (13.2.49) is K in (13.2.43) which, therfore, is not singular anywhere.
In this way, as far as its solution of interest is concerned, the ODE in (13.2.46) has a removable singularity.
On the other hand, as seen in the first relation in (13.2.44), this does not make H nonsingular at $\epsilon = 0$, in the sense of (13.2.23). And in fact, H fails to be even $\mathcal{C}^1$-smooth in a neighbourhood of $\epsilon = 0$.

3. Standard Reduction of Nonautonomous ODEs to Autonomous ODEs

Here, for convenience and for setting up the notation which will be used later, we shortly present a well known procedure in systems theory, which allows the reduction of nonautonomous ODEs to autonomous ones.

First, let us consider the autonomous nonlinear system of ODEs

$$(13.3.1) \qquad \begin{aligned} &\frac{d}{d\epsilon} Y(\epsilon) = F(Y(\epsilon)), \quad \epsilon \in \mathbf{R} \\ &Y(\epsilon_0) = y_0 \in \mathbf{R}^l \end{aligned}$$

where $F \in \mathcal{C}^\infty(\mathbf{R}^l, \mathbf{R}^l)$, $\epsilon_0 \in \mathbf{R}$, $y_0 \in \mathbf{R}^l$ are given, while $Y \in \mathcal{C}^\infty(\mathbf{R}, \mathbf{R}^l)$ is the solution. For simplicity, we shall assume that the unique solution Y exists for all $\epsilon \in \mathbf{R}$. Then the associated $\mathcal{C}^\infty$-smooth evolution operator E is given by

$$(13.3.2) \qquad \begin{aligned} &\mathbf{R} \ni \epsilon \longmapsto E(\epsilon) : \mathbf{R}^l \longrightarrow \mathbf{R}^l \\ &E(\epsilon - \epsilon_0)(y) = Y(\epsilon), \quad \epsilon_0, \epsilon \in \mathbf{R}, \quad y \in \mathbf{R}^l \end{aligned}$$

and it obviously defines the one dimensional Lie group action on $\mathbf{R}^l$

$$(13.3.3) \qquad \mathbf{R} \times \mathbf{R}^l \ni (\epsilon, y) \longmapsto E(\epsilon)(y) \in \mathbf{R}^l$$

In particular, the evolution operator E has the group properties

$$(13.3.4) \qquad \begin{aligned} &E(0) = id_{\mathbf{R}^l} \\ &E(\epsilon + \eta) = E(\epsilon)E(\eta), \quad \epsilon, \eta \in \mathbf{R} \end{aligned}$$

in other words, we have the group homomorphism

(13.3.5) $\mathbf{R} \ni \epsilon \longmapsto E(\epsilon) \in \mathcal{D}iff^{\infty}(\mathbf{R}^l)$

Let us now consider the nonautonomous nonlinear system of ODEs

$$(13.3.6)\qquad \begin{aligned} &\frac{d}{d\epsilon}Y(\epsilon) = F(\epsilon, Y(\epsilon)), \quad \epsilon \in \mathbf{R} \\ &Y(\epsilon_0) = y_0 \in \mathbf{R}^l \end{aligned}$$

where $F \in \mathcal{C}^{\infty}(\mathbf{R}^{l+1}, \mathbf{R}^l)$, $\epsilon_0 \in \mathbf{R}$, $y_0 \in \mathbf{R}^l$ are given, and the unique solution $Y \in \mathcal{C}^{\infty}(\mathbf{R}, \mathbf{R}^l)$ is supposed to exist for all $\epsilon \in \mathbf{R}$. This time, the associated nonautonomous evolution operator $E \in \mathcal{C}^{\infty}(\mathbf{R}^2 \times \mathbf{R}^l, \mathbf{R}^l)$ has the property

$$(13.3.7)\qquad \begin{aligned} &\mathbf{R} \times \mathbf{R} \ni (\epsilon_0, \epsilon) \longmapsto E(\epsilon_0, \epsilon) : \mathbf{R}^l \longrightarrow \mathbf{R}^l \\ &E(\epsilon_0, \epsilon)(y_0) = Y(\epsilon), \quad \epsilon_0, \epsilon \in \mathbf{R}, \; y_0 \in \mathbf{R}^l \end{aligned}$$

However, it is important to note that if we solve the nonautonomous ODE system in (13.3.6) with the initial conditions $y_0 \in \mathbf{R}^l$, for a given fixed $\epsilon_0 \in \mathbf{R}$ only, then (13.3.7) will in general not give the full information on E, and do so in an explicit manner. This is unlike the case with autonomous ODEs, see (13.3.2). Nevertheless, as seen in (13.3.22), the full information on the nonautonomous evolution operator E can be recovered from (13.3.7), without having to solve the ODE system (13.3.6) for evey other $\epsilon_0 \in \mathbf{R}$, but only by solving some additional algebraic equations. Indeed, this becomes possible, once we transform the nonautonomous ODE system in (13.3.6) into the autonomous ODE system (13.3.10), and then we use the corresponding autonomous evolution operator E_A.

In the case (13.3.6), (13.3.7), the relations (13.3.4) take the following more general, nonautonomous form of group property, namely

$$(13.3.8)\qquad \begin{aligned} &E(\epsilon, \epsilon) = id_{\mathbf{R}^l}, \quad \epsilon \in \mathbf{R} \\ &E(\eta, \theta)E(\epsilon, \eta) = E(\epsilon, \theta), \quad \epsilon, \eta, \theta \in \mathbf{R} \end{aligned}$$

Now we recall the standard way the nonautonomous system (13.3.6) can be reduced to an autonomous one, such as for instance in (13.3.1).
Let us augment the function $Y \in \mathcal{C}^{\infty}(\mathbf{R}, \mathbf{R}^l)$ in (13.3.6) to the function $Y_A \in \mathcal{C}^{\infty}(\mathbf{R}, \mathbf{R}^{l+1})$, given by

(13.3.9) $Y_A(\epsilon) = (\epsilon, Y(\epsilon)), \quad \epsilon \in \mathbf{R}$

then clearly, the l-dimensional nonautonomous ODE system (13.3.6) is equivalent with the $l+1$-dimensional autonomous ODE system

$$(13.3.10)\qquad \begin{aligned} &\frac{d}{d\epsilon}Y_A(\epsilon) = (1, F(Y_A(\epsilon))), \quad \epsilon \in \mathbf{R} \\ &Y_A(\epsilon_0) = (\epsilon_0, y_0) \in \mathbf{R}^{l+1} \end{aligned}$$

Two facts should be noted here.
First, the nonautonomous infinitesimal generator F in (13.3.6), becomes associated in (13.3.10) with the autonomous infinitesimal generator $F_A \in \mathcal{C}^\infty(\mathbf{R}^{l+1}, \mathbf{R}^{l+1})$ defined by

$$(13.3.11) \quad F_A(y_A) = (1, F(y_A)), \quad y_A \in \mathbf{R}^{l+1}$$

Second, in the initial condition (ϵ_0, y_0) in (13.3.10), the first coordinate $\epsilon_0 \in \mathbf{R}$ must be the same with that in the left hand term. Clearly, this is a direct consequence of the definition of Y_A in (13.3.9).

Let now $E_A \in \mathcal{C}^\infty(\mathbf{R} \times \mathbf{R}^{l+1}, \mathbf{R}^{l+1})$ be the autonomous evolution operator associated with (13.3.10), then

$$(13.3.12) \quad E_A(\epsilon - \epsilon_0)(\epsilon_0, y_0) = Y_A(\epsilon), \quad \epsilon_0, \epsilon \in \mathbf{R}, \; y_0 \in \mathbf{R}^l$$

and it is easy to see that it can be decomposed as follows

$$(13.3.13) \quad E_A(\eta)(\epsilon, y) = (E_{A1}(\eta)(\epsilon, y), E_{A2}(\eta)(\epsilon, y)), \quad \epsilon, \eta \in \mathbf{R}, \; y \in \mathbf{R}^l$$

with the $\mathcal{C}^\infty$-smooth functions

$$(13.3.14) \quad \begin{aligned} &\mathbf{R} \ni \eta \longmapsto E_{A1}(\eta) : \mathbf{R}^{l+1} \longrightarrow \mathbf{R} \\ &\mathbf{R} \ni \eta \longmapsto E_{A2}(\eta) : \mathbf{R}^{l+1} \longrightarrow \mathbf{R}^l \end{aligned}$$

In view of (13.3.7), (13.3.9), (13.3.12), (13.3.13), it follows that, for $\epsilon_0, \epsilon \in \mathbf{R}$, $y_0 \in \mathbf{R}^l$, we have

$$(13.3.15) \quad E(\epsilon_0, \epsilon)(y_0) = Y(\epsilon) = E_{A2}(\epsilon - \epsilon_0)(\epsilon_0, y_0)$$

Also, E_A satisfies the group properties corresponding to the autonomous case (13.3.4), namely

$$(13.3.16) \quad \begin{aligned} &E_A(0) = id_{\mathbf{R}^{l+1}} \\ &E_A(\epsilon + \eta) = E_A(\eta) E_A(\epsilon), \quad \epsilon, \eta \in \mathbf{R} \end{aligned}$$

which result in the group homomorphism

$$(13.3.17) \quad \mathbf{R} \ni \epsilon \longmapsto E_A(\epsilon) \in \mathcal{D}iff^\infty(\mathbf{R}^{l+1})$$

Therefore, for $\epsilon, \eta, \theta \in \mathbf{R}$ and $y \in \mathbf{R}^l$, we obtain the relations

$$(13.3.18) \quad \begin{aligned} &E_{A1}(\theta + \eta)(\epsilon, y) = E_{A1}(\theta)(E_{A1}(\eta)(\epsilon, y), E_{A2}(\eta)(\epsilon, y)) \\ &E_{A2}(\theta + \eta)(\epsilon, y) = E_{A2}(\theta)(E_{A1}(\eta)(\epsilon, y), E_{A2}(\eta)(\epsilon, y)) \end{aligned}$$

And in view of (13.3.7) - (13.3.12), we obtain for $\epsilon, \eta \in \mathbf{R}$, $y \in \mathbf{R}^l$

$$E_{A1}(\eta)(\epsilon, y) = \epsilon + \eta$$

(13.3.19)

$$E_{A2}(\eta)(\epsilon, y) = E(\epsilon, \epsilon + \eta)(y)$$

In particulár, we can check for E the nonautonomous form of the semigroup property (13.3.8). Namely, for $\epsilon, \eta, \theta \in \mathbf{R},\ y \in \mathbf{R}^l$, we have

(13.3.20)

$$\begin{aligned} E(\epsilon, \epsilon + \eta + \theta)(y) &= \\ &= E_{A2}(\eta + \theta)(\epsilon, y) = \\ &= E_{A2}(\theta)(\epsilon + \eta, E_{A2}(\eta)(\epsilon, y)) = \\ &= E_{A2}(\theta)(\epsilon + \eta, E(\epsilon, \epsilon + \eta)(y)) = \\ &= E(\epsilon + \eta, \epsilon + \eta + \theta)E(\epsilon, \epsilon + \eta)(y) \end{aligned}$$

Let us show now how we can recover the full nonautonomous evolution operator E in (13.3.8), from one of its particular cases in (13.3.7), corresponding to a certain $\epsilon_0 \in \mathbf{R}$ fixed. For that, we shall use the autonomous extension E_A of E.

Indeed, (13.3.15) gives

(13.3.21) $\quad E(\epsilon, \eta)(y) = E_{A2}(\eta, \epsilon)(\epsilon, y), \quad \epsilon,\ \eta \in \mathbf{R},\ y \in \mathbf{R}^l$

We fix now $\epsilon_0 \in \mathbf{R}$. Then given any $\epsilon \in \mathbf{R}$, if we can compute the correspondence

$$\mathbf{R}^l \ni y \longmapsto y_* \in \mathbf{R}^l$$

where y_* is the solution of the algebraic equation

$$E(\epsilon_0, \epsilon)(y_*) = y$$

it is clear that we thus obtain $(\epsilon, y) = E_A(\epsilon - \epsilon_0)(\epsilon_0, y_0)$, hence

$$E(\epsilon, \eta)(y) = E_{A2}(\eta - \epsilon)E_A(\epsilon - \epsilon_0)(\epsilon_0, y_*) = E(\epsilon_0, \eta)(y_*)$$

Therefore

(13.3.22)

$$E(\epsilon, \eta)(y) = E(\epsilon_0, \eta)(y_*), \quad \epsilon, \eta \in \mathbf{R},\ y \in \mathbf{R}^l$$

provided that y_* is a solution of $E(\epsilon_0, \epsilon)(y_*) = y$

We shall also assume that

(13.3.23) $\quad y_*$ depends $\mathcal{C}^\infty$ – smoothly on $\epsilon,\ y$

Let us illustrate the above with a simple example. We consider the nonautonomous ODE

$$\frac{d}{d\epsilon}Y(\epsilon) = 2\epsilon, \quad \epsilon \in \mathbf{R}$$

$$Y(\epsilon_0) = y_0 \in \mathbf{R}$$

for which obviously

$$Y(\epsilon) = E(\epsilon_0, \epsilon)(y_0) = E_{A2}(\epsilon - \epsilon_0)(\epsilon_0, y_0) = \epsilon^2 - \epsilon_0^2 + y_0$$

$$Y_A(\epsilon) = (\epsilon, E_{A2}(\epsilon - \epsilon_0)(\epsilon_0, y_0))$$

$$E_{A1}(\eta)(\epsilon, y) = \epsilon + \eta$$

$$E_{A2}(\eta)(\epsilon, y) = E(\epsilon, \epsilon + \eta)(y) = \eta^2 + 2\eta\epsilon + y$$

We now check the group properties (13.3.15). First, we note that

$$E_A(0) = id_{\mathbf{R}^{l+1}}$$

results immediately from the expressions of E_{A1} and E_{A2} above. Then

$$E_A(\theta)E_A(\eta) = E_A(\eta + \theta), \quad \eta, \theta \in \mathbf{R}$$

holds since the previous relations give $E_A(\eta)(\epsilon, y) = (\epsilon + \eta, \eta^2 + 2\eta\epsilon + y)$, hence $E_A(\theta)E_A(\eta)(\epsilon, y) = E_A(\theta)(\epsilon+\eta, \eta^2+2\eta\epsilon+y) = (\epsilon+\eta+\theta, \theta^2+2\theta(\epsilon+\eta)+\eta^2+2\eta\epsilon+y)$, while $E_A(\eta + \theta)(\epsilon, y) = (\epsilon + \eta + \theta, (\eta + \theta)^2 + 2(\eta + \theta)\epsilon + y)$.

In other words, the nonautonomous version (13.3.8) of the group property works as follows, for $\epsilon, \eta, \theta \geq 0,\ y \in M$, we have

$$E(\epsilon, \eta)(y) = \eta^2 - \epsilon^2 + y$$

hence clearly $E(\epsilon, \epsilon)(y) = y$, while

$$E(\eta, \theta)E(\epsilon, \eta)(y) = \theta^2 - \eta^2 + (\eta^2 - \epsilon^2 + y) = \theta^2 - \epsilon^2 + y = E(\epsilon, \theta)(y)$$

The autonomous version, which corresponds to the nonautonomous one, according to (13.3.9) - (13.3.22), will have the group property (13.3.16) working as follows. For $\epsilon, \eta, \theta \geq 0,\ y \in M$, we have

$$E_A(\eta)(\epsilon, y) = (\epsilon + \eta, \eta^2 + 2\eta\epsilon + y)$$

thus obviously $E_A(0)(\epsilon, y) = (\epsilon, y)$, while

$$E_A(\theta)E_A(\eta)(\epsilon, y) = (\epsilon + \eta + \theta, \theta^2 + 2\theta(\epsilon + \eta) + \eta^2 + 2\eta\epsilon + y) =$$

$$(\epsilon + \eta + \theta, (\eta + \theta)^2 + 2(\eta + \theta)\epsilon + y) = E_A(\eta + \theta)(\epsilon, y)$$

Finally, in the case of this example, the relation (13.3.22) works as follows. Given a fixed $\epsilon_0 \in \mathbf{R}$, the algebraic equation

$$\epsilon^2 - \epsilon_0^2 + y_* = y$$

can obviously be solved in y_* for every $\epsilon, y \in \mathbf{R}$, and it gives $y_* = y - \epsilon^2 + \epsilon_0^2$, which also satisfies (13.3.23), therefore

$$E(\epsilon, \eta)(y) = E(\epsilon_0, \eta)(y_*) = \eta^2 - \epsilon_0^2 + y_* = \eta^2 - \epsilon_0^2 + y$$

4. Applications to Earlier Examples of Actions

We can now return to the examples in section 2, and show the way in which the respective actions can be associated with *genuine* Lie semigroup actions, by applying to them the method of reduction in section 3.
However, first we have to note that the results in section 3, where the ODEs are defined on the whole of **R**, have to be adapted since, this time, with the examples in section 2, we will no longer have $\epsilon \in \mathbf{R}$, but only $\epsilon \in (0, \infty)$, or at most $\epsilon \in [0, \infty)$, see (13.2.10), (13.2.25) - (13.2.27), (13.2.36) - (13.2.38). As noted in section 2, this restriction of the domain of ϵ is due to the noremovable singularites of the repsective ODEs, and in particular, of their solutions of interest.

Concerning the autonomous case (13.3.1) - (13.3.5), such a restriction of the domain of ϵ means that the respective evolution operator E in (13.3.2), will only be defined for $\epsilon \in [0, \infty)$, and instead of (13.3.4), will have the semigroup property

$$(13.4.1) \quad E(\epsilon + \eta) = E(\eta)E(\epsilon), \quad \epsilon,\ \eta \in [0, \infty)$$

In particular, $E(\epsilon)$, with $\epsilon \in (0, \infty)$, may fail to be invertible, since $E(-\epsilon)$ need not exist, and thus, we could not always obtain from (13.3.4) the relations $E(\epsilon)E(-\epsilon) = E(-\epsilon)E(\epsilon) = id_{\mathbf{R}^l}$.

In the nonautonomous case (13.3.6) - (13.3.8), the evolution operator E in (13.3.7) will only be defined for $\epsilon_0,\ \epsilon \in [0, \infty)$, and will have the following semigroup version of property (13.3.8)

$$(13.4.2) \quad E(\theta, \eta)E(\epsilon, \eta) = E(\epsilon, \theta), \quad \epsilon,\ \eta,\ \theta \in [0, \infty)$$

This however allows for the existence of its ***inverses***, since we have for any $\epsilon,\ \eta \in [0, \infty)$

$$(13.4.3) \quad \begin{array}{l} E(\epsilon, \eta)E(\eta, \epsilon) = E(\eta, \eta) = id_{\mathbf{R}^l} \\ E(\eta, \epsilon)E(\epsilon, \eta) = E(\epsilon, \epsilon) = id_{\mathbf{R}^l} \end{array}$$

Let us now transform the thus restricted nonautonomous version of (13.3.6) - (13.3.8) into the corresponding restricted autonomous version of (13.3.9) - (13.3.22). Then clearly, for $\epsilon,\ \eta \in [0, \infty),\ y \in \mathbf{R}^l$, we obtain

$$(13.4.4) \quad E_A(\eta)(\epsilon, y) = (\epsilon + \eta, E(\epsilon, \epsilon + \eta)(y))$$

or equivalently

$$(13.4.5) \quad E_A(\eta - \epsilon)(\epsilon, y) = (\eta, E(\epsilon, \eta)(y)), \quad 0 \le \epsilon \le \eta < \infty$$

Therefore, in the case of such singular nonautonomous ODEs and of their corresponding evolution operators E, after the transformation into the autonomous case, except for the trivial situation of $\eta = \epsilon$, we need no longer be able to benefit from the existence of the inverses in (13.4.3), when we deal with the associated autonomous evolution operator E_A.

This is precisely at the basis of our construction of *genuine* Lie semigroup actions. And in particular, this is how we shall associate such genuine Lie semigroup actions with the examples in section 2.

Before going further, let us note that in the case of both nonautonomous examples in section 2, we do have the corresponding versions of (13.4.2), namely

(13.4.6) $\quad E(\eta, \theta)E(\epsilon, \eta) = E(\epsilon, \theta), \quad \epsilon,\ \eta,\ \theta \in [0, \infty)$

Indeed, for the example in (13.2.10) - (13.2.20), this follows from the fact that H in (13.2.10) is a solution on $(0, \infty)$ of (13.2.16) - (13.2.20), while in addition, see (13.2.21), H is such that $H(0, y) = y$, for all $y \in M$, and also $H \in \mathcal{C}^0([0, \infty) \times M, M)$.
A similar argument will apply to the more general example in (13.2.24) - (13.2.34).

In view of (13.4.6), we obtain for the example in (13.2.10) - (13.2.20), the relation

(13.4.7) $\quad E(0, \epsilon)(y) = y + \sqrt{\epsilon} y^2, \quad \epsilon \in [0, \infty),\ y \in M$

while for the example in (13.2.24) - (13.2.34), we shall have

(13.4.8) $\quad E(0, \epsilon)(y) = (1 - \alpha(\epsilon))y + \alpha(\epsilon)f(y), \quad \epsilon \in [0, \infty),\ y \in \mathbf{R}^l$

These two relations will help us in fully computing the respective nonautonomous evolution operators E for the mentioned examples.
Indeed, let us determine $E(\epsilon, \eta)(y)$ for the first example, and do so for all $\epsilon,\ \eta \in (0, \infty)$, $y \in M = \mathbf{R}$. From (13.4.6) we have

$$E(\epsilon, \eta)(y) = E(0, \eta)E(\epsilon, 0)(y) = E(0, \eta)(E(0, \epsilon))^{-1}(y)$$

Hence, proceeding for $\epsilon_0 = 0$ as in (13.3.22), let us assume that

$$(E(0, \epsilon))^{-1}(y) = y_* \in M$$

then clearly

$$y = E(0, \epsilon)(y_*) = y_* + \sqrt{\epsilon} y_*^2$$

and we note the important consequence that

$$\lim_{\epsilon \to 0} y_* = y \in M$$

provided that y_* is bounded.

Now computing y_* from the above quadratic equation, we obtain

$$y_* = \frac{-1 \pm \sqrt{1+4\sqrt{\epsilon}y}}{2\sqrt{\epsilon}}, \quad \epsilon \in (0,\infty),\ y \in M,\ 1+4\sqrt{\epsilon}y \geq 0$$

and in order to secure the above limit, it follows that we must choose

$$y_* = \frac{-1 + \sqrt{1+4\sqrt{\epsilon}y}}{2\sqrt{\epsilon}} = \frac{2y}{1+\sqrt{1+4\sqrt{\epsilon}y}}, \quad \epsilon \in (0,\infty),\ y \in M,\ 1+4\sqrt{\epsilon}y \geq 0$$

In this way (13.3.23) is satisfied, while

$$E(\epsilon,\eta)(y) = E(0,\eta)(y_*) =$$

$$(13.4.9) \qquad = \frac{2y}{1+\sqrt{1+4\sqrt{\epsilon}y}} + \sqrt{\eta}\,\frac{4y^2}{\left(1+\sqrt{1+4\sqrt{\epsilon}y}\right)^2}$$

$$\epsilon,\ \eta \in (0,\infty),\ y \in M,\ 1+4\sqrt{\epsilon}y \geq 0$$

and we have obtained the full expression of the nonautonomous evolution operator E for the first example in section 2.

The *genuine* Lie semigroup of actions generated by the example in (13.2.10) - (13.2.20) will now be given by the autonomous operator E_A, which corresponds to E above, according to (13.3.9), namely, see also (13.4.4)

$$(13.4.10) \qquad E_A(\eta)(\epsilon,y) = (\epsilon+\eta, E(\epsilon,\epsilon+\eta)(y))$$

$$\epsilon,\ \eta \in [0,\infty),\ y \in M,\ 1+4\sqrt{\epsilon}y \geq 0$$

The genuine Lie semigroup actions generated by the example in (13.2.24) - (13.2.34) can be obtained in a similar manner, provided that one can solve in $y_0 \in \mathbf{R}^l$ the corresponding algebraic equations

$$y = (1-\alpha(\epsilon))y_0 + \alpha(\epsilon)f(y_0)$$

for given $(\epsilon, y) \in (0,\infty) \times \mathbf{R}^l$.

Similar arguments apply to the class of examples in (13.2.35) - (13.2.41).
We end this section with an example which shows that condition (13.2.30) is not necessary, in order to obtain genuine Lie group actions by using the above method. Indeed, let $M = \mathbf{R}$ and $f \in \mathcal{C}^\infty(M,M)$ be given by

$$(13.4.11) \qquad f(y) = \frac{1}{y^2+1}, \quad y \in M$$

We define $H \in \mathcal{C}^0([0,\infty) \times M, M) \cap \mathcal{C}^\infty((0,\infty) \times M, M)$ such that

$$(13.4.12) \quad H(\epsilon, y) = (1-\sqrt{\epsilon})y + \sqrt{\epsilon} f(y) = (1-\sqrt{\epsilon})y + \frac{\sqrt{\epsilon}}{y^2+1}, \quad \epsilon \in [0,\infty),\ y \in M$$

Then it follows that

$$(13.4.13) \quad \begin{aligned} & H(1,\ .\ ) = f \notin \mathcal{D}iff^\infty(M) \\ & H(\epsilon,\ .\ ) \in \mathcal{D}iff^\infty(M) \quad \text{for } \epsilon \in [0, 4/9) \cup (4, \infty) \end{aligned}$$

However, it is obvious that the above procedure applied to the examples in section 2, is equally applicable to (13.4.11) - (13.4.13), and again, it will lead to genuine Lie semigroup actions.

5. Continuity, Smoothness and Domains of Action

In view of sections 2 - 4, it appears that one possible appropriate concept of genuine Lie semigroup actions is that given by the evolution operators E_A of the autonomous singular ODEs, which are associated in the above standard manner with the nonautonomous singular ODEs in (13.2.21) - (13.2.23).
In other words, such genuine Lie semigroup actions on suitable subsets $\bar{M}$ in Euclidean spaces are given by mappings

$$(13.5.1) \quad E_A \in (\mathcal{C}^0([0,\infty) \times \bar{M}, \bar{M}) \cap \mathcal{C}^\infty((0,\infty) \times \bar{M}, \bar{M})) \setminus \mathcal{C}^1([0,\infty) \times \bar{M}, \bar{M})$$

With respect to the domains of action of such genuine Lie semigroups, as constructed in sections 2 - 4, we have to note the following.
We have started with certain open subsets M in Euclidean spaces, and with singular actions, see (13.2.10), (13.2.27), (13.2.36)

$$(13.5.2) \quad H \in (\mathcal{C}^0([0,\infty) \times M, M) \cap \mathcal{C}^\infty((0,\infty) \times M, M)) \setminus \mathcal{C}^1([0,\infty) \times M, M)$$

which clearly were not any kind of semigroup actions on the respective open subsets M.
Then, we found nonautonomous singular ODEs which were satisfied by these H, and associated with them in the standard manner autonomous singular ODEs.
Finally, the evolution operators E_A of these associated autonomous ODEs gave us the genuine Lie semigroup actions.

However, such an E_A is *no longer* acting on M, but on

$$(13.5.3) \quad \bar{M} = [0,\infty) \times M$$

In particular, this method can be used in an obvious manner, in order to incorporate the actions, respectively semisymmetries, in the examples in section 1 into genuine Lie semigroup actions.
Indeed, the mapping (13.1.3), for instance, which satisfies (13.2.24), can be incorporated into the procedure in (13.2.25) - (13.2.34), and then, as shown in section 4, can be dealt with by the method in section 3.

6. Remark on Singularities

Let us note that in order to obtain genuine Lie semigroups, we can use *milder* forms of singularities than those in section 2. For instance, instead of H given by (13.2.10), let us consider it defined as follows

$$(13.6.1) \quad H(\epsilon, y) = y + \epsilon y^2, \quad \epsilon \in \mathbf{R},\ y \in M = \mathbf{R}$$

Then (13.2.11) is replaced with

$$(13.6.2) \quad \begin{aligned} & H(0, y) = y, \quad y \in M \\ & H(\epsilon, \ .\ ) \notin \mathcal{D}iff^{\infty}(M), \quad \epsilon \in \mathbf{R} \setminus \{0\} \\ & H \in \mathcal{C}^{\infty}(\mathbf{R} \times M, M) \end{aligned}$$

while the action in (13.2.12) extends to the following action on M

$$(13.6.3) \quad \mathbf{R} \times M \ni (\epsilon, y) \longmapsto H(\epsilon, y) \in M$$

which again cannot be part of a group or local group action on M, since in view of (13.6.2), $H(\epsilon, \ .\ )$, with $\epsilon \in \mathbf{R} \setminus \{0\}$, is not a $\mathcal{C}^{\infty}$-smooth diffeomorphism of M.

Proceeding now in a manner similar with that in (13.2.13) - (13.2.20), it follows that H, as a function of $\epsilon \in \mathbf{R}$, and for any given fixed $y \in M$, will satisfy the ODE

$$(13.6.4) \quad \begin{aligned} & \frac{\partial}{\partial \epsilon} Y(\epsilon) = \frac{2Y(\epsilon)^2}{1 + 2\epsilon Y(\epsilon) + \sqrt{1 + 4\epsilon Y(\epsilon)}}, \\ & \qquad\qquad \epsilon \in \mathbf{R},\ 1 + 4\epsilon Y(\epsilon) \geq 0 \end{aligned}$$

if

$$(13.6.5) \quad 1 + 2\epsilon y \geq 0$$

while it will satisfy the ODE

$$(13.6.6) \quad \begin{aligned} & \frac{\partial}{\partial \epsilon} Y(\epsilon) = \frac{1 + 2\epsilon Y(\epsilon) + \sqrt{1 + 4\epsilon Y(\epsilon)}}{2\epsilon^2}, \\ & \qquad\qquad \epsilon \in \mathbf{R} \setminus \{0\},\ 1 + 4\epsilon Y(\epsilon) \geq 0 \end{aligned}$$

if

$$(13.6.7) \quad 1 + 2\epsilon y \leq 0$$

Furthermore, both these ODEs will for H specified above be associated with the initial condition

$$(13.6.8) \quad \lim_{\epsilon \to 0} Y(\epsilon) = y \in M$$

We can note in the above example (13.6.1) - (13.6.8) that in the case of (13.6.5), the corresponding ODE in (13.6.4) which is satisfied by H, seen as a function of $\epsilon \in \mathbf{R}$, and with $y \in M$ fixed, is *not* singular at $\epsilon = 0$.
Neverhteless, the respective solution H, for $\epsilon \in \mathbf{R} \setminus \{0\}$, still *cannot* be part of a group or local group of transformations on M, in view of (13.6.2).

7. Evolution PDEs and Genuine Lie Semigroups

In this section, as a possible alternative to section 2, we give an indication about another way one may try to generate genuine Lie semigroup actions. This possible way is suggested by the well established literature on solving initial value problems for evolution PDEs through the associated semigroups of operators acting on the respective intial values. The early basic result in the case of linear evolution PDEs is the celebrated Hille-Yoshida theorem, which was follwed by a rather large body of more recent results, including nonlinear developments, see for instance Pazy and the references cited there.

The important point to note here is that, in general, the solutions of the initial value problems for evolution PDEs will be given by semigroups, rather than groups, of such operators. A well known class of evolution PDEs for which, typically, one can only obtain such semigroups of operators is that of parabolic equations.
The idea suggested in this section is to try to use such semigroups acting on initial values, in order to generate the genuine Lie semigroups which are the object of our interest in the present chapter.

For simplicity, we shall again consider the case when $M = \mathbf{R}^l$ is the open set on which we want to define a genuine Lie semigroup action.
Let us therefore take any evolution PDE of the form

$$(13.7.1) \quad D_t U(t,x) = T(x,D)U(t,x), \quad t \in [0,\infty),\ x \in M$$

where $U \in \mathcal{C}^\infty([0,\infty) \times M, \mathbf{R})$ is the unknown function, while $T(x,D)$ is a partial differential operator in x alone, such as in (1.1), (1.4).
Associated with (13.7.1), we consider the initial value problem

$$(13.7.2) \quad U(0,x) = f(x), \quad x \in M$$

where f belongs to a suitable class of functions, namely

$$(13.7.3) \quad f \in \mathcal{F}(M) \subseteq \mathcal{C}^\infty(M, \mathbf{R})$$

We shall assume that there exists a semigroup of operators acting on the typically *infinite* dimensional vector space $\mathcal{F}(M)$, namely

$$(13.7.4) \quad [0,\infty) \ni t \longmapsto E(t) : \mathcal{F}(M) \to \mathcal{F}(M)$$

such that, given any $f \in \mathcal{F}(M)$, if we define

$$(13.7.5) \quad U(t,x) = (E(t)f)(x), \quad t \in [0,\infty),\ x \in M$$

then U is a solution of (13.7.1), (13.7.2).

So far, we have been moving within the well established framework of the mentioned literature, provided that we work with suitable evolution equations (13.7.1) and initial values (13.7.2).

Now the idea which is the subject of this section is to make the following further assumption. Namely, suppose that there exists a family of functions

$$(13.7.6) \quad V_{a,b} \in \mathcal{F}(M), \quad a \in M, \ b \in B$$

where $B \subseteq \mathbf{R}^k$ is a suitable open subset, such that

$$(13.7.7) \quad E(t)V_{a,b} = V_{\alpha(t,a,b),\beta(t,a,b)}, \quad t \in [0,\infty), \ a \in M, \ b \in B$$

for appropriate functions

$$(13.7.8) \quad \alpha \in \mathcal{C}^\infty([0,\infty) \times M \times B, M), \ \beta \in \mathcal{C}^\infty([0,\infty) \times M \times B, B)$$

In this case, the assumed semigroup property

$$(13.7.9) \quad E(s)E(t) = E(t+s), \quad t,s \in [0,\infty)$$

together with (13.7.6), (13.7.7) will result in

$$(13.7.10) \quad \begin{aligned} \alpha(t+s,a,b) &= \alpha(s,\alpha(t,a,b),\beta(t,a,b)) \\ \beta(t+s,a,b) &= \beta(s,\alpha(t,a,b),\beta(t,a,b)) \end{aligned}$$

for $t,s \in [0,\infty), \ a \in M, \ b \in B$.

It follows that if, for instance

$$(13.7.11) \quad \beta(t,a,b) = b, \quad t \in [0,\infty), \ a \in M, \ b \in B$$

then (13.7.10) yields

$$(13.7.12) \quad \alpha(t+s,a,b) = \alpha(s,\alpha(t,a,b),b), \quad t,s \in [0,\infty), \ a \in M, \ b \in B$$

hence for any fixed $b \in B$, the mapping

$$(13.7.13) \quad [0,\infty) \times M \ni (t,a) \longmapsto \alpha(t,a,b) \in M$$

is a semigroup action on M, and it is a priori not impossible that it may indeed be a *genuine Lie semigroup* action.

It is easy to see that the assumptions (13.7.6) - (13.7.8) can be satisfied by a large variety of *soliton* solutions, for instance. Indeed, for simplicity, let us consider the one dimensional case, when $M = \mathbf{R}$, and assume that (13.7.1) has a soliton solution

$$(13.7.14) \quad U(t,x) = W(x-ct), \quad t \in [0,\infty), \ x \in M$$

for $c \in \mathbf{R}$. In this case (13.7.5) becomes

$$(E(t)W)(x) = W(x - ct), \quad t \in [0, \infty), \; x \in M \tag{13.7.15}$$

For further more detailed illustration, let us consider the Burgers equation

$$U_t(t,x) + U(t,x)U_x(t,x) = \mu U_{xx}(t,x), \quad t, x \in \mathbf{R} \tag{13.7.16}$$

where $\mu \in (0, \infty)$. This equation has a soliton solution given by

$$U(t,x) = c - \sqrt{c^2+d}\ \tanh\left(\frac{\sqrt{c^2+d}}{2\mu}\,(x - x_0 - ct)\right), \quad t, x \in \mathbf{R} \tag{13.7.17}$$

where $x_0, c, d \in \mathbf{R}$ are arbitrary fixed, and $c^2 + d > 0$. For this soliton it is clear that (13.7.6) - (13.7.8), (13.7.11) - (13.7.13) are satisfied, if we take

$$a = x_0 \in M = \mathbf{R}, \quad b = (c,d) \in B \subset \mathbf{R}^2 \tag{13.7.18}$$

where $B = \{(c,d) \in \mathbf{R}^2 \mid c^2 + d > 0\}$, in which case

$$\begin{aligned} \alpha(t,a,b) &= x_0 + ct \\ \beta(t,a,b) &= b \end{aligned} \tag{13.7.19}$$

for $t \in [0,\infty)$, $a \in M$, $b \in B$.

8. Other Instances of Semigroups of Actions

In a private correspondence, P J Olver mentioned further instances in which semigroups of actions appear in a natural way. Not all of them, however, need be genuine Lie semigroup actions.

A first such example happens in the framework of (1.2), (1.3), when a certain subset $S \subset M$ is given, and we are interested in the set of actions which invariate it, namely

$$G_S = \{\, g \in G \mid gS \subseteq S \,\}$$

For instance, let

$$S = (-1,1) \times \mathbf{R} \subset \mathbf{R}^2 = M$$

and $G = (\ (0,\infty),\ .\ )$ be the usual Lie group, which is supposed to act on M according to

$$G \times M \ni (g,(x,y)) \longmapsto (gx, y) \in M$$

Then clearly

$$G_S = (0, 1]$$

which is a semigroup action. However, G_S is not a genuine Lie semigroup action, since it is a subsemigroup of the Lie group action G.

A second example is given by ODEs with inequality constraints. Let us, for instance, consider the differential equation

$$\frac{d}{dx}U(x) = 0, \quad x \in \mathbf{R}$$

with the inequality constraint

$$U(x) > 0, \quad x \in \mathbf{R}$$

Then the Lie group actions

$$(x, u) \longmapsto (x, u + c)$$

are symmetries, only if $c \geq 0$. Needless to say, in view of a large class of applications, such as control theory or differential games, for instance, where ODEs with inequality constraints play a crucial role, the study of semigroups of symmetries of such equations can present a special interest.

Another class of examples, this time related to PDEs, is given by generalised symmetries, see Olver [1, chap. 5]. Indeed, the evolution PDEs governing the flow of a generalised symmetry often only define a semigroup. A good example is the symmetry

$$V = U_{xx}\partial_U$$

which is a symmetry of any linear constant coefficient PDE. Its flow is

$$U_t = U_{xx}$$

Appendix

Let be given any $\mathcal{C}^\infty$-smooth mapping

(A.1) $\quad X \xrightarrow{f} Y$

where $X, Y \subseteq \mathbf{R}^n$ are nonvoid, open. We recall that f is a *local diffeomorphism* , if and only if

$$\text{(A.2)} \qquad \begin{array}{l} \forall\, x \in X : \\ \exists\, \Delta \subseteq X \ \text{ open neighbourhood of } x : \\ \quad \Delta \xrightarrow{f} f(\Delta) \ \text{ is a } \mathcal{C}^\infty\text{-smooth diffeomorphism} \end{array}$$

In this case $f(\Delta) \subseteq Y$ is nonvoid, open, hence so is $f(X) \subseteq Y$.

In general, a $\mathcal{C}^\infty$-smooth local diffeomorphism need *not* be injective, and hence, a diffeomorphism, even if it is surjective, see Parthasarathy, Pinchuk.

Given any f in (A.1), let $J_f \ : \ X \longrightarrow \mathbf{R}$ be the Jacobian of f. Then

(A.3) $\quad X_f = \{x \in X \mid J_f(x) \neq 0\}$

is open, and in case $X_f \neq \phi$, we obtain from the inverse function theorem the $\mathcal{C}^\infty$-smooth surjective local diffeomorphism

(A.4) $\quad X_f \xrightarrow{f} f(X_f)$

where $f(X_f)$ is open.

If $X_f \neq \phi$, let

$$\text{(A.5)} \qquad \mathcal{X}_f \ = \ \left\{ \Delta \subseteq X_f \;\middle|\; \begin{array}{l} \text{1) } \Delta \text{ nonvoid, open} \\ \text{2) } \Delta \xrightarrow{f} f(\Delta) \\ \quad\ \text{is a } \mathcal{C}^\infty\text{-smooth diffeomorphism} \end{array} \right\}$$

Then clearly, $f(\Delta)$ is nonvoid and open, for $\Delta \in \mathcal{X}_f$. And it is easy to see that $\mathcal{X}_f$ is *chain complete.* Indeed, let $\Delta_i \in \mathcal{X}_f$, with $i \in I$, be a family totally ordered by inclusion. Then

$$\Delta = \bigcup_{i \in I} \Delta_i \subseteq X_f$$

is again nonvoid, open. Furthermore

$$\Delta \xrightarrow{f} f(\Delta)$$

is obviously a $\mathcal{C}^\infty$-smooth surjective local diffeomorphism. However f is also injective, since given any $x, x' \in \Delta$, there exists $i \in I$, such that $x, x' \in \Delta_i$. And therefore $x \neq x'$ will imply $f(x) \neq f(x')$. It follows that

$$\Delta \xrightarrow{f} f(\Delta)$$

is a $\mathcal{C}^\infty$-smooth diffeomorphism, which means that $\Delta \in \mathcal{X}_f$. In this way, Zorn's lemma yields the nonvoid set of nonvoid open subsets in X

$$\text{(A.6)} \qquad \bar{\mathcal{X}}_f = \{\bar{\Delta} \in \mathcal{X}_f \mid \bar{\Delta} \text{ maximal in } \mathcal{X}_f\}$$

and we have

$$\begin{aligned} & \forall\ \Delta \in \mathcal{X}_f\ : \\ \text{(A.7)} \qquad & \exists\ \bar{\Delta} \in \bar{\mathcal{X}}_f\ : \\ & \quad \Delta \subseteq \bar{\Delta} \end{aligned}$$

Moreover, (A.4), (A.7) yield

$$\text{(A.8)} \qquad X_f = \cup\, \bar{\Delta}$$

where the union is taken over all $\bar{\Delta} \in \bar{\mathcal{X}}_f$. Also, it follows that

$$\begin{aligned} & \forall\ \bar{\Delta} \in \bar{\mathcal{X}}_f\ : \\ \text{(A.9)} \qquad & \forall\ \Delta \subseteq X_f \backslash \bar{\Delta}, \text{ nonvoid, open:} \\ & \quad f\,(\Delta) \cap f(\bar{\Delta}) \neq \phi \end{aligned}$$

Indeed, let us take $x \in \Delta$. Then (A.4) yields $x \in \Delta' \subseteq \Delta$, with $\Delta' \in \mathcal{X}_f$. Let us assume that $f(\Delta) \cap f(\bar{\Delta}) = \phi$, then in particular $f(\Delta') \cap f(\bar{\Delta}) = \phi$. However, this implies that $\bar{\Delta}' = \bar{\Delta} \cup \Delta' \in \mathcal{X}_f$, thus contradicting the maximality of $\bar{\Delta}$. A consequence of (A.9) is the property

$$\begin{aligned} & \forall\ \bar{\Delta} \in \bar{\mathcal{X}}_f\ : \\ \text{(A.10)} \qquad & \\ & \quad f(X_f \backslash \bar{\Delta}) \subseteq \text{cl } f(\bar{\Delta}) \end{aligned}$$

where cl denotes the usual topological closure in $\mathbf{R}^n$. To prove it, let $x \in X_f \backslash \bar{\Delta}$. In case $x \in \text{cl } \bar{\Delta}$, the continuity of f gives $f(x) \in \text{cl } f(\bar{\Delta})$. Otherwise, there exists an open neighbourhood Δ of x, with $\Delta \cap \bar{\Delta} = \phi$. In view of (A.4), we can assume that $\Delta \in \mathcal{X}_f$. Hence for every $x \in \Delta' \subseteq \Delta, \Delta'$ open, we obtain from (A.9) that $f(\Delta') \cap f(\bar{\Delta}) \neq \phi$. But since $\Delta \in \mathcal{X}_f$, it follows that $f(\Delta')$ is an arbitrary open neighbourhood of $f(x)$, hence (A.10) obtains. Clearly, (A.10) implies

$$\forall\ \bar{\Delta} \in \bar{\mathcal{X}}_f\ :$$

(A.11)
$$f(X_f) \subseteq \text{cl } f(\bar{\Delta})$$

Remark A

As we deal with an arbitrary $\mathcal{C}^\infty$-smooth function f in (A.1), for which the mapping in (A.4) is in general only a $\mathcal{C}^\infty$ -smooth surjective local diffeomorphism, that is, in the terms of (A.3), (A.5), we need *not* have, see (A.32)

$$X_f \in \mathcal{X}_f$$

it follows that the above property (A.11) gives a rather *strong* statement about the size of any maximal $\bar{\Delta} \in \mathcal{X}_f$. Namely, when seen from the *range* of f, the difference between $f(X_f)$ and the smaller $f(\bar{\Delta})$ is *topologically minimal*, since the former is always contained in the closure of the latter.

In view of the particular situation that both the domain X and range Y of f in (A.1) are of the same dimension n, it follows, Kahn, that in case $X_f \neq \phi$, the mapping, see (A.4)

$$X_f \xrightarrow{f} f(X_f)$$

will be an *immersion*. Further, for every $\Delta \in \mathcal{X}_f$, the mapping, see (A.5)

$$\Delta \xrightarrow{f} f(\Delta)$$

will be an *embedding*. Finally, if $\bar{\Delta} \in \bar{\mathcal{X}}_f$, then the mapping, see (A.6)

$$\bar{\Delta} \xrightarrow{f} f(\bar{\Delta})$$

will be a *maximal* embedding.

□

In view of the above, with each $\mathcal{C}^\infty$-smooth mapping $X \xrightarrow{f} Y$ in (A.1), for which $X_f \neq \phi$, we shall associate the family of $\mathcal{C}^\infty$-smooth diffeomorphisms

(A.12)
$$f|_{diff} = \left(\Delta \xrightarrow{f|_\Delta} \Delta' \;\middle|\; \begin{array}{l} 1)\;\; \Delta \in \mathcal{X}_f \\ 2)\;\; \Delta' = f(\Delta) \end{array} \right)$$

Clearly, each $f|_\Delta : \Delta \longrightarrow f(\Delta)$, with $\Delta \in \mathcal{X}_f$, is a $\mathcal{C}^\infty$-smooth diffeomorphism, and the family (A.12) contains *all* such $\mathcal{C}^\infty$-smooth diffeomorphism which - as restrictions to open sets in X - can be associated with a given arbitrary f in (A.1). In view of (A.9) - (A.11), the family (A.12) *cannot* in general be indexed by the nonvoid open subsets $\Delta' = f(\Delta) \subseteq f(X_f)$, since the mapping

(A.13)
$$\mathcal{X}_f \ni \Delta \longmapsto \Delta' = f(\Delta) \subseteq f(X_f)$$

is *not* necessarily injective.

Recapitulating the above, our interest is in the mapping, see (A.1), (A.12) and (2.11)

(A.14) $\quad f \longmapsto f|_{diff}$

in the case when

(A.15) $\quad X \xrightarrow{f} Y$

is a $\mathcal{C}^\infty$-smooth mapping for which $X_f \neq \phi$, see (A.3), and which in general may nevertheless fail to be a $\mathcal{C}^\infty$-smooth diffeomorphism as well. The problem we face here is that, in general, f in (A.15) need *not* be $\mathcal{C}^\infty$-smooth invertible. In particular, our problem is that, given another $\mathcal{C}^\infty$-smooth mapping $X \xrightarrow{h} Z$, with Z an open subset of some Euclidean space, we would like to construct a commutative diagram

(A.16)

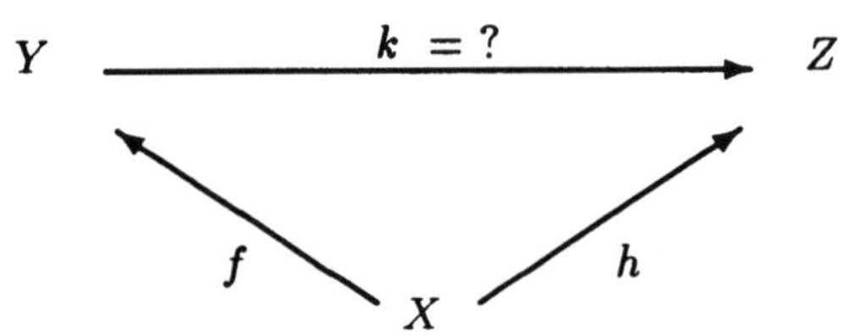

even if f^{-1} does *not* exist as a $\mathcal{C}^\infty$-smooth function, and thus we *cannot* simply define $k = h \circ f^{-1}$ in the usual set theoretic way for functions.

Here, as a help, comes in the mapping $f \longmapsto f|_{diff}$ in (A.14) which, in general, is the *best one can do* towards solving the problem in (A.16). Namely, instead of (A.16), we can construct the *family* of commutative diagrams

(A.17)

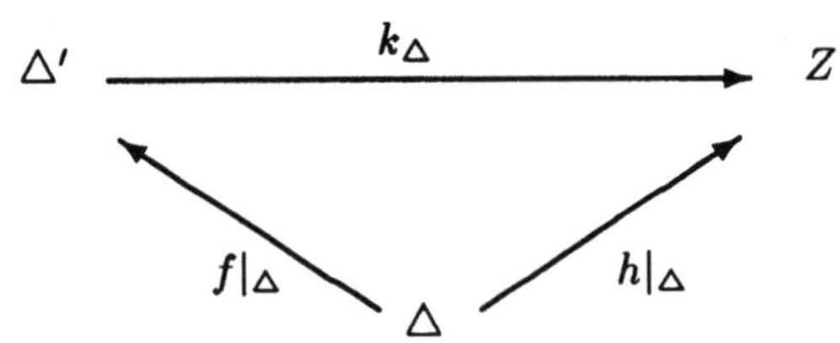

where $\Delta \in \mathcal{X}_f$, while

(A.18) $\quad \Delta' = f(\Delta), \quad k_\Delta = h|_\Delta \circ (f|_\Delta)^{-1}$

Clearly

(A.19) $\quad \cup\Delta = X_f \ , \quad \cup\Delta' = f(X_f)$

where the unions are taken over all $\Delta \in \mathcal{X}_f$, that is, those defining the commutative diagrams (A.17). And as noted above for the family in (A.12), similarly here, the family of commutative diagrams (A.17) cannot be indexed by the open sets $\Delta' = f(\Delta) \subseteq f(X_f)$, since the mapping (A.13) need not be injective.

We conclude the above by introducing the notation, see (A.16), (A.17)

$$\text{(A.20)} \quad h \circ f^{-1} = (h|_\Delta \circ (f|_\Delta)^{-1} | \Delta \in \mathcal{X}_f)$$

and in view of (A.13), it follows that for certain nonvoid open subsets $\Delta' \subseteq f(X_f)$, we may have *more than one* mapping

$$\text{(A.21)} \quad h|_\Delta \circ (f|_\Delta)^{-1} \ : \ \Delta' \longrightarrow Z$$

Now we return to the issue of the size of the open set $f(X_f)$ in (A.3), (A.4), and do so in a different manner than in (A.11) or in Remark A. Namely, this time, we shall look at the singularities of f in (A.1) and use Sard's theorem, Kahn, in order to obtain an estimate of

$$f(X) \setminus f(X_f)$$

Therefore, for f in (A.1), let us denote its set of *singularities* by

$$\text{(A.22)} \quad \sum_f = \{x \in X | J_f(x) = 0\} = X \setminus X_f$$

which will obviously be closed in X. Then Sard's theorem gives that

$$\text{(A.23)} \quad \text{mes } f(\textstyle\sum_f) = 0$$

where mes denotes the Lebesgue measure on $\mathbf{R}^n$. Now an elementary set theoretic argument yields

$$f(X) \setminus f(X_f) \subseteq f(\textstyle\sum_f)$$

hence (A.23) results in

$$\text{(A.24)} \quad \text{mes } (f(X) \setminus f(X_f)) = 0$$

Furthermore, (A.4) implies that $f(X) \setminus f(X_f)$ is closed in $f(X)$, hence (A.24) gives

(A.25) $f(X) \setminus f(X_f)$ is closed, nowhere dense in the interior of $f(X)$

The results in (A.11), Remark A, (A.24) and (A.25) can now be brought together in the following manner

$$(\text{A.26})\qquad \begin{aligned}&\forall\ \bar{\Delta}\in\bar{\mathcal{X}}_f\ :\\ &\quad 1)\ f(\bar{\Delta})\subseteq f(X_f)\subseteq f(X)\cap \text{cl}\ f(\bar{\Delta})\\ &\quad 2)\ \text{mes}\ (f(X)\backslash f(X_f))=0\\ &\quad 3)\ f(X)\backslash f(X_f)\ \text{is closed in}\ f(X)\\ &\quad 4)\ f(X)\backslash f(X_f)\ \text{is closed, nowhere dense}\\ &\qquad \text{in the interior of}\ f(X)\end{aligned}$$

We note that the index set $\mathcal{X}_f$, which appears in both families (A.12) and (A.20), will in general only have the following two properties, namely, a *hereditary* property

$$(\text{A.27})\qquad \begin{aligned}&\forall\ \Delta_1\ \subseteq\ \Delta\ \in\ \mathcal{X}_f\ :\\ &\quad \Delta_1\ \text{nonvoid, open}\ \Longrightarrow \Delta_1\ \in\ \mathcal{X}_f\end{aligned}$$

and a *chain completeness* property

$$(\text{A.28})\qquad \begin{aligned}&\forall\ \Delta_i\ \in\ \mathcal{X}_f\ ,\text{with}\ i\ \in\ I\ :\\ &\quad (\Delta_i | i\in I)\ \text{totally ordered}\ \Longrightarrow \bigcup_{i\in I}\Delta_i\in\mathcal{X}_f\end{aligned}$$

In view of (A.27), (A.28) and (A.7), (A.8), it is clear that $f|_{diff}$ in (A.12) contains *redundancies*. Namely, it can be fully reconstituted from the knowledge of its terms corresponding to *maximal* open, nonvoid sets Δ only. In this way, upon convenience, we shall on occasion replace (A.12) with its maximal subfamily given by

$$(\text{A.29})\qquad f|_{diff}\ =\ \left(\bar{\Delta}\xrightarrow{f|_{\bar{\Delta}}}\bar{\Delta}'\ \middle|\ \begin{array}{l}1)\ \bar{\Delta}\in\bar{\mathcal{X}}_f\\ 2)\ \bar{\Delta}'=f(\bar{\Delta})\end{array}\right)$$

Similarly, $h\circ f^{-1}$ in (A.20) will upon convenience be replaced with its redundance free version

$$(\text{A.30})\qquad h\circ f^{-1}\ =\ (\ h|_{\bar{\Delta}}\circ(f|_{\bar{\Delta}})^{-1}\ |\ \bar{\Delta}\in\bar{\mathcal{X}}_f\ )$$

Given an arbitrary $\mathcal{C}^\infty$-smooth mapping

$$(\text{A.31})\qquad X\xrightarrow{f}Y$$

as in (A.1), the particular case, when the mapping in (A.4), namely

$$(\text{A.32})\qquad X_f\xrightarrow{f}f(X_f)$$

happens to be a $\mathcal{C}^\infty$-smooth diffeomorphism, has the following natural characterisation

(A.33) $\quad \bar{\mathcal{X}}_f = \{ X_f \}$

or equivalently

(A.34) $\quad \mathcal{X}_f = \{\Delta \subseteq X_f \,|\, \Delta \text{ nonvoid, open}\}$

Clearly, (A.33) means that $\bar{\mathcal{X}}_f$ contains a *largest*, hence also maximal, open set, and then, in view of (A.8), this must be X_f. In such a case, the equivalence with (A.34) follows easily.

The point of interest to us is that in case (A.32) - (A.34) hold, we can *further simplify* the objects of our main concern, given in (A.12) and (A.20), respectively in (A.29) and (A.30). Indeed, in view of (A.34), all the information in (A.12) is already contained in its *largest* term, which corresponds to $\Delta = X_f$, since the rest of the terms can be obtained simply by *restriction* to all nonvoid, open subsets in X_f. Therefore, in such a case, we shall further simplify (A.12) and (A.29), by writing them as the following *one single term family*

(A.35) $\quad f|_{diff} = (X_f \xrightarrow{f} f(X_f))$

Similarly, we shall further simplify (A.20) and (A.30) to the form

(A.36) $\quad h \circ f^{-1} = (h|_{X_f} \circ (f|_{X_f})^{-1})$

In this way, at last, in the yet more particular case, when f in (A.1), (A.31) is itself a $\mathcal{C}^\infty$-smooth *diffeomorphism*, that is

(A.37) $\quad X_f = X \in \mathcal{X}_f \text{ and } \bar{\mathcal{X}}_f = \{ X \}$

then we are able to *recover* the classical variants of (A.12) and (A.29), namely

(A.38) $\quad f|_{diff} = f : X \longrightarrow Y$

as well as of (A.20) and (A.30) respectively

(A.39) $\quad h \circ f^{-1} : X \longrightarrow Z$

Let us here note that (A.1) - (A.21) and (A.27) - (A.30) may under suitable conditions be extended to infinite dimensional cases.

Finally, we note that the construction in (A.1), (A.3) - (A.12) may recall the concept of *covering space*, Munkres, which however proves to be somewhat more particular. Indeed, if $X \xrightarrow{f} Y$ is a covering for Y, then f is a local homeomorphism, although the converse need not hold. On the other hand, there is a well known link between homeomorphisms and coverings, Berger.

In Olver, [2], any $\mathcal{C}^\infty$-smooth surjective local diffeomorphism $X \xrightarrow{f} Y$ is called a covering of Y by X.

Example A

We shall illustrate (A.1) - (A.11) in the case of a rather simple function. Nevertheless, we shall be able to note the potential for *complexity*, when less simple functions may be considered.
Let $X = Y = \mathbf{R}$ and

$$X \ni x \xrightarrow{f} x^3 - 3x \in Y \tag{A.40}$$

then (A.3) gives

$$X_f = \mathbf{R} \setminus \{-1, 1\}, \quad f(X_f) = \mathbf{R} \tag{A.41}$$

In order to determine $\bar{\mathcal{X}}_f$ in (A.6), let us note that

$$f(-\infty) = -\infty \nearrow f(-1) = 2 \searrow f(1) = -2 \nearrow f(\infty) = \infty$$

and

$$f(-2) = -2, \quad f(2) = 2$$

Let us now take any $b \in [-2, 2]$ and denote by

$$-2 \leq a_1 \leq -1 \leq a_2 \leq 1 \leq a_3 \leq 2$$

the corresponding roots of the equation $f(x) = b$. We note that

$$b = -2 \Leftrightarrow a_1 = -2 \Leftrightarrow a_2 = a_3 = 1$$

$$-2 < b < 2 \Leftrightarrow -2 < a_1 < -1 < a_2 < 1 < a_3 < 2$$

$$b = 2 \Leftrightarrow a_1 = a_2 = -1 \Leftrightarrow a_3 = 2$$

We define now the following nonvoid open subsets in $X = \mathbf{R}$

$$\begin{aligned} \bar{\Delta}_{1,b} &= (-\infty, a_1) \cup (-1, a_2) \cup (2, \infty) \\ \bar{\Delta}_{2,b} &= (-\infty, -2) \cup (a_2, 1) \cup (a_3, \infty) \end{aligned} \tag{A.42}$$

then in particular, we obtain

$$\begin{aligned} \bar{\Delta}_{1,-2} &= (-\infty, -2) \cup (-1, 1) \cup (2, \infty) \\ \bar{\Delta}_{2,-2} &= (-\infty, -2) \cup (1, \infty) \\ \bar{\Delta}_{1,2} &= (-\infty, -1) \cup (2, \infty) \\ \bar{\Delta}_{2,2} &= (-\infty, -2) \cup (-1, 1) \cup (1, \infty) \end{aligned} \tag{A.43}$$

In this way, it follows easily that

$$(A.44) \qquad \bar{\mathcal{X}}_f \supseteq \{\, \bar{\Delta}_{1,b} \mid b \in [-2,2] \,\} \bigcup \{\, \bar{\Delta}_{2,b} \mid b \in [-2,2] \,\}$$

However, $\bar{\mathcal{X}}_f$ contains many more maximal nonvoid open subsets in $\mathbf{R}$. For instance, for $b \in [-2,2]$, let us define

$$(A.45) \qquad \bar{\Delta}_{3,b} = (-\infty,-2) \cup (-1,a_2) \cup (1,a_3) \cup (2,\infty)$$

then clearly

$$(A.46) \qquad \bar{\mathcal{X}}_f \supseteq \{\, \bar{\Delta}_{3,b} \mid b \in [-2,2] \,\}$$

and this is obviously not the end of the list of maximal nonvoid open subsets in $\mathbf{R}$ which belong to $\bar{\mathcal{X}}_f$.

We note that f in (A.40) is a function of one single real variable, and it only has one local maximum, at $x = -1$, and one local minimum, at $x = 1$, and it has no other extrema. Nevertheless, as seen in (A.44) and (A.45), $\bar{\mathcal{X}}_f$ contains a large variety of nonvoid open maximal subsets in $X = \mathbf{R}$.
Needless to say, in the case of functions f in (A.1) of more than one real variable, and with a larger number of extrema, the structure of $\bar{\mathcal{X}}_f$ can become rather complex.

We note that in the above example, if given, for instance, $x = -3$, then there are *infinitely* many different $\bar{\Delta} \in \bar{\mathcal{X}}_f$, such that $x \in \bar{\Delta}$.

Finally, related to (A.10) and (A.11), we note that

$$\begin{aligned} \bar{\Delta} &= (-\infty,-1) \cup (2,\infty) \in \bar{\mathcal{X}}_f \\ f(\bar{\Delta}) &= \mathbf{R} \setminus \{2\} \\ X_f &= (-1,1) \cup (1,2] \\ f(X_f \setminus \bar{\Delta}) &= (-2,2] \end{aligned}$$

therefore

$$f(X_f \setminus \bar{\Delta}) \not\subseteq f(\bar{\Delta}), \quad \text{hence} \quad f(X_f) \not\subseteq f(\bar{\Delta})$$

Bibliography

[1] J. Aczel, The state of the second part of Hilbert's fifth problem. Bull. AMS, Vol. 20, No. 2, April 1989, 153-163

[2] P.C. Aichelburg and H. Balasin [1], A note on the symmetries of the gravitational field of a massless particle. Class. Quant. Grav. 11, L121-L123, 1994

[3] P.C. Aichelburg and H. Balasin [2], Symmetries of PP waves with distributional profile. Class. Quant. Grav. 13, 723-730, 1996

[4] P.C. Aichelburg and H. Balasin [3], Generalised symmetries of impulsive gravitational waves. Class. Quant. Grav. 14, A31-A42, 1997

[5] S. Albeverio, Z. Haba, F. Russo, Trivial solutions for a non-linear two-space dimensional wave equation perturbed by space-time white noise. Stochastics and Stochastic Reports, Vol. 56, 1996, 127-160

[6] R.L. Anderson and N.H. Ibragimov, Lie-Backlund Transformations in Applications. SIAM, Philadelphia, 1979

[7] H. Balasin [1], Colombeau's generalised functions on arbitrary manifolds. Preprint GR-QC/9610017, 1996

[8] H. Balasin [2], Geodesics for impulsive gravitational waves and the multiplication of distributions. Class. Quant. Grav. 14, 455-462, 1997

[9] H. Balasin [3], Distributional energy momentum tensor of the extended Kerr geometry. Class. Quant. Grav. 14, 3353-3362, 1997

[10] H. Balasin and H. Nachbagauer, The energy momentum tensor of a black hole or what curves the Schwazschild geometry ? Class. Quant. Grav. 10, 2271-2278, 1993

[11] H.A. Biagioni, A Nonlinear Theory of Generalized Functions. Springer Lecture Notes in Mathematics, Vol. 1421, Springer, New York, 1990

[12] M.S. Berger, Nonlinearity and Functional Analysis. Acad. Press, New York, 1977

[13] R.J. Blattner, On geometric quantization. In (Eds. S.I. Anderson and H-D Doebner) Nonlinear Partial Differential Operators and Quantum Procedures. Springer Lecture Notes in Mathematics, Vol. 1037, 209-241. Springer, New York, 1983

[14] G.W. Bluman and S. Kumei, Symmetries and Differential Equations. Springer, New York, 1989

[15] J-L Brylinski, Loop Spaces, Characteristic Classes and Geometric Quantization. Birkhauser, Boston, 1993

[16] F.E. Browder (Ed), Mathematical Developments Arising from Hilbert Problems. Proceedings of Symposia in Pure Mathematics, Vol. XXVIII, AMS Providence, 1976

[17] C. Chevalley, Theory of Lie Groups I. Princeton Univ. Press, 1946

[18] C.J.S. Clarke [1], The Analysis of Space-Time Singularities. Cambridge Univ. Press, 1993

[19] C.J.S. Clarke [2], Singularities : boundaries or internal points. In (Ed. P.S. Joshi) Singularities, Black Holes and Cosmic Censorship, pp. 24-32, IUCCA, Bombay, 1996

[20] C.J.S. Clarke [3], Generalised hyberbolicity in singular space-times. Preprint GR-QC/9702033, 1997

[21] C.J.S. Clarke, J.A. Vickers and J.P.Wilson, Generalised functions and distributional curvature of cosmic strings. Class. Quant. Grav. 13, 1996, 2485-2498

[22] J-F Colombeau, Elementary Introduction to New Generalized Functions. North-Holland Mathematics Studies, Vol. 113, Amsterdam, 1985

[23] F.J. Dyson, Missed opportunities. Bull. AMS, Vol. 78, No. 5, Sept. 1972, 635-652

[24] L.C. Evans, Weak Convergence Methods for Nonlinear PDEs. Conference Board of Mathematical Sciences, No. 74, AMS, Providence, 1990

[25] R. Geroch and J.Traschen, Strings and other distributional sources in general relativity. Phys. Rev. D 36 (4), 1987, 1017-1031

[26] L. Gillman and M. Jerison, Rings of Continuous Functions. Van Nostrand, New York, 1960

[27] M. Grosser, G. Hoermann, M. Kunzinger and M. Oberguggenberger (Eds.), Nonlinear Theory of Generalized Functions. Pitman Research Notes in Mathematics, Longman, Harlow, London, 1998

[28] M. Heller [1], Algebraic foundations of the theory of differential spaces. Demonstratio Math., 24, 1991, 349-364

[29] M. Heller [2], Einstein algebras and general relativity. Intern.J. Theor.Physics, 31, 1992, 277-288

[30] R. Hermann and M.B. Oberguggenberger, Generalized functions, calculus of variations, and nonlinear ODEs. Preprint, Univ. Innsbruck, 1995

[31] D. Hilbert, Mathematical Problems. Bull. AMS, Vol. 8, 1902, 437-479 (reprinted in F.E. Browder)

[32] J. Hilgert and K.H. Neeb, Lie Semigroups and their Applications. Springer Lecture Notes in Mathematics, Vol. 1552, Heidelberg, 1993

[33] J. Hilgert, K.H. Hofmann and J.D. Lawson, Lie Groups, Convex Cones, and Semigroups. Oxford Univ. Press, Oxford, 1989

[34] J. Hirschfeld, The nonstandard treatment of Hilbert's fifth problem. Trans. AMS, Vol. 321, No. 1, Sept. 1990, 379-400

[35] N.H. Ibragimov, CRC Handbook of Lie Group Analysis of Differential Equations, Vol. 1, CRC Press, Boca Raton, 1994

[36] A. Kaneko, Introduction to Hyperfunctions. Kluwer, Dordrecht, 1988

[37] D.W. Kahn, Introduction to Global Analysis. Acad. Press, New York, 1980.

[38] A.A. Kirillov [1], Elements of the Theory of Representations. Springer, New York, 1976

[39] A.A. Kirillov [2], Geometric quantization. In (Eds. V.I. Arnold and S.P. Novikov) Dynamical Systems IV. Symplectic Geometry and its Applications, 137-172. Springer, New York, 1990

[40] A. Kriegl and P.W. Michor, The Convenient Setting of Global Analysis. AMS Mathematical Surveys and Monographs, Vol. 53, 1997

[41] M. Kunzinger, Lie transformation groups in Colombeau algebras. Doctoral Thesis, Univ. Vienna, 1996

[42] M. Kunzinger and R. Steinbauer, A rigorous solution concept for geodesic equations in impulsive gravitational waves. Preprint 1998

[43] M. Kunzinger, M. Grosser, R. Steinbauer and M.B. Oberguggenberger, Geometric Theory of Generalized Functions. To appear in Kluwer, Dordrecht, The Netherlands

[44] S. Lie, Theorie der Transformationsgruppen. Vol. 3, B.G. Teubner, Leipzig, 1893

[45] Z. Lozanov Crvenković, S. Pilipović, Gaussian generalized random processes. Univ. Novom Sadu Zb. Rad. Prirod.-Mat. Fak. Ser. Mat., Vol. 25, No. 2, 1995, 57-60

[46] A. Mallios [1], Geometry of Vector Sheaves. An Axiomatic Approach to Differential Geometry, Vols. I (Chapts. 1-5), II (Chapts. 6-11). Kluwer, Dordrecht, 1997

[47] A. Mallios [2], On an axiomatic treatment of differntial geometry via vbector sheaves. Applications. Intern. Plaza Math. Japonica (invited paper in press)

[48] A. Mallios [3], The de Rham-Kahler complex of the Gel'fand sheaf of a topological algebra. J.Math.Anal.Appl., 175, 1993, 143-168

[49] A. Mallios [4], On an abstract for of Weil's integrality theorem. Note di Mat., 12, 1992, 167-202 (invited paper)

[50] A. Mallios [5], On an axiomatic approach to geometric prequantization : A classification scheme a la Kostant-Souriau-Kirillov. J. Math. Sci. (former J.Soviet Math.) (invited paper in press)

[51] A. Mallios [6], Abstract differential geometry, general relativity, and singularities. (to appear)

[52] A. Mallios and E.E. Rosinger, Abstract differential geometry, differential algebras of generalized functions, and de Rham cohomology (to appear)

[53] C. Martias [1], Stochastic integration on generalized function spaces and its applications. Stochastics and Stochastics Reports, Vol. 57, 1996, 289-301

[54] C. Martias [2], Extension du calcul d'Itô aux distributions de Schwartz. C. R. Acad. Sci. Paris, Sér. I, Vol. 321, 1995, 331-334

[55] P.W. Michor, Gauge Theory for Fiber Bundles. Monographs and Textbooks in Physical Science, Bibliopolis, Napoli, 1991

[56] D. Montgomery and L. Zippin, Topological Transformation Groups. Intersci.Publ., New York, 1955

[57] J.R. Munkres, Topology, A First Course. Prentice-Hall, New Jersey, 1975

[58] M.B. Oberguggenberger [1], Multiplication of Distributions and Applications to PDEs. Pitman Research Notes in Mathematics, Vol. 259. Longman, Harlow, 1992

[59] M.B. Oberguggenberger [2], Generalized functions and stochastic processes. In: (Eds.) E. Bolthausen, M. Dozzi, F. Russo, Seminar on Stochastic Analysis, Random Fields and Applications. Birkhäuser-Verlag, Basel, 1995, 215-229

[60] M.B. Oberguggenberger and F. Russo [1], Nonlinear stochastic wave equations. Proc. Conf. Generalized Functions Novi Sad 1996. In: Integral Transforms and Special Functions, Vol. 6, 1997, 58-70

[61] M.B. Oberguggenberger and F. Russo [2], Nonlinear SPDEs: Colombeau solutions and pathwise limits. In: (Eds.) L. Decreusefond, J. Gjerde, B. Øksendal, A.S. Üstünel, Stochastic Analysis and Related Topics. Birkhäuser-Verlag, Boston, 1998, 319-332

[62] M.B. Oberguggenberger and E.E. Rosinger, Solution of Continuous Nonlinear PDEs through Order Completion. North-Holland Mathematics Studies, Vol. 181. North-Holland, Amsterdam, 1994

[63] P.J. Olver [1], Applications of Lie Groups to Differential Equations. Springer, New York, 1986

[64] P.J. Olver [2], Equivalence, Invariants and Symmetry. Cambridge Univ. Press, 1995

[65] P.J. Olver [3], Nonassociative local Lie groups. J. Lie Theory, 6, 1996, 23-51

[66] J.C. Oxtoby, Measure and Category. Springer, New York, 1971

[67] T. Parthasarathy, On Global Univalence Theorems. Springer Lecture Notes in Mathematics, Vol. 977, Springer, New York, 1983.

[68] A. Pazy, Semigroups of Linear Operators and Applications to PDEs. Applied Mathematical Sciences, Vol. 44, Springer, New York, 1983

[69] S. Pinchuk, A counterexample to the strong real Jacobian conjecture. Math.Z., Vol. 217, 1994, 1-4

[70] C. Rogers and W.F. Shadwick, Backlund Transformations and Their Applications. Acad. Press, New York, 1982

[71] E.E. Rosinger [1], Embedding of the $\mathcal{D}'$ distributions into pseudo-topological algebras. Stud. Cerc. Math., Vol. 18, No. 5, 1966, 687-729.

[72] E.E. Rosinger [2], Pseudotopological spaces, the embedding of the $\mathcal{D}'$ distributions into algebras. Stud. Cerc. Math., Vol. 20, No. 4, 1968, 553-582.

[73] E.E. Rosinger [3], Division of Distributions. Pacif.J. Math., Vol. 66, No. 1, 1976, 257-263

[74] E.E. Rosinger [4], Nonsymmetric Dirac distributions in scattering theory. In Springer Lecture Notes in Mathematics, Vol. 564, 391-399, Springer, New York, 1976

[75] E.E. Rosinger [5], Distributions and Nonlinear Partial Differential Equations. Springer Lectures Notes in Mathematics, Vol. 684, Springer, New York, 1978.

[76] E.E. Rosinger [6], Nonlinear Partial Differential Equations, Sequential and Weak Solutions, North Holland Mathematics Studies, Vol. 44, Amsterdam, 1980.

[77] E.E. Rosinger [7], Generalized Solutions of Nonlinear Partial Differential Equations. North Holland Mathematics Studies, Vol. 146, Amsterdam, 1987.

[78] E.E. Rosinger [8], Nonlinear Partial Differential Equations, An Algebraic View of Generalized Solutions. North Holland Mathematics Studies, Vol. 164, Amsterdam, 1990.

[79] E.E. Rosinger [9], Global version of the Cauchy-Kovalevskaia theorem for nonlinear PDEs. Acta Appl. Math., Vol. 21, 1990, 331–343.

[80] E.E. Rosinger [10], Characterization for the solvability of nonlinear PDEs, Trans. AMS, Vol. 330, 1992, 203–225.

[81] E.E. Rosinger [11], Nonprojectable Lie Symmetries of nonlinear PDEs and a negative solution to Hilbert's fifth problem. In (Eds. N.H. Ibragimov and F.M. Mahomed) Modern Group Analysis VI, Proceedings of the International Conference in the New South Africa, Johannesburg, January 1996, 21-30. New Age Inter. Publ., New Delhi, 1997

[82] E.E. Rosinger [12], Arbitrary Lie group actions on global generalized solutions of nonlinear PDEs and an answer to Hilbert's fifth problem (manuscript)

[83] E.E. Rosinger and M. Rudolph, Group invariance of global generalised solutions of nonlinear PDEs : A Dedekind order completion method. Lie Groups and their Applications, Vol. 1, No. 1, July-August 1994, 203-215

[84] E.E. Rosinger and E.Y. Walus [1], Group invariance of generalized solutions obtained through the algebraic method. Nonlinearity, Vol. 7, 1994, 837-859

[85] E.E. Rosinger and E.Y. Walus [2], Group invariance of global generalised solutions of nonlinear PDEs in nowhere dense algebras. Lie Groups and their Applications, Vol. 1, No. 1, July-August 1994, 216-225

[86] W.Rudin, Principles of Mathematical Analysis. McGraw-Hill, New York, 1964

[87] F. Russo, Colombeau generalized functions and stochastic analysis. In: (Eds.) A.I. Cardoso, M. de Faria, J. Potthoff, R. Sénéor, L. Streit, Analysis and Applications in Physics. Kluwer, Dordrecht, 1994, 329-349

[88] R. Steinbauer [1], The ultrarelativistic Reissner-Nordstrom field in the Colombeau algebra. J. Math. Phys. 38 (3), 1997, 164-1622

[89] R. Steinbauer [2], Geodesics and the geodesic deviation for impulsive gravitational waves. Preprint ESI/505, to appear in J. Math. Phys.

[90] F. Schur, Ueber den analytischen Charakter der eine endliche continuirliche Transformationsgruppe darstellenden Funtionen. Math.Ann., 41, 1893, 509-538

[91] D.J. Simms and N.M.J. Woodhouse, Lectures on Geometric Quantization. Springer Lecture Notes in Physics, Vol. 53, Springer, New York, 1976

[92] J. Smoller, Shock Waves and Reaction-Diffusion Equations. Springer, New York, 1983

[93] J-M Souriau [1], Structures des Systemes Dynamiques. Dunod, Paris, 1970

[94] J-M Souriau [2], Groupes differentiels. In Differential Geometric Methods in Mathematical Physics. Springer Lecture Notes In Mathematics, Vol. 863, 91-128. Springer, New York, 1980

[95] R.C. Walker, The Stone-Cech Compatification. Springer, Heidelberg, 1974

[96] E. Vassiliou [1], On Mallios A-connections as connections on principal sheaves. Note di Mat., 14, 1994, 237-249

[97] E. Vassiliou [2], Connections on principal sheaves. (to appear in New Developments in Differential Geometry, Kluwer, 459-484)

[98] E. Vassiliou [3], Topological algebras and abstract differential geometry. (to appear in J. Math. Sci.)

[99] E. Vassiliou [4], Flat principal sheaves. Preprint Univ. of Athens, Grece

[100] C. von Westenholz, Differential Forms in Mathematical Physics. North-Holland, Amsterdam, 1981

[101] J.P. Wilson, Distributional curvature of the time dependent cosmic strings. Class. Quant. Grav. 14, 3337-3351, 1997

[102] A. Weinstein, Grupoids : unifying internal and external symmetry, a tour through some examples. Notices of the AMS, Vol. 43, No. 7, July 1996, 744-752

[103] C.T. Yang, Hilbert's fifth problem and related problems on transformation groups. In F.E. Browder (Ed), Mathematical Developments Arising from Hilbert Problems, Proceedings of Symposia in Pure Mathematics, Vol. XXVIII, 142-146, AMS, Providence, 1976

Index

Other *Mathematics and Its Applications* titles of interest:

V.A. Vassiliev: *Ramified Integrals, Singularities and Lacunas*. 1995, 289 pp. ISBN 0-7923-3193-1

N.Ja. Vilenkin and A.U. Klimyk: *Representation of Lie Groups and Special Functions*. Recent Advances. 1995, 497 pp. ISBN 0-7923-3210-5

Yu. A. Mitropolsky and A.K. Lopatin: *Nonlinear Mechanics, Groups and Symmetry*. 1995, 388 pp. ISBN 0-7923-3339-X

R.P. Agarwal and P.Y.H. Pang: *Opial Inequalities with Applications in Differential and Difference Equations*. 1995, 393 pp. ISBN 0-7923-3365-9

A.G. Kusraev and S.S. Kutateladze: *Subdifferentials: Theory and Applications*. 1995, 408 pp. ISBN 0-7923-3389-6

M. Cheng, D.-G. Deng, S. Gong and C.-C. Yang (eds.): *Harmonic Analysis in China*. 1995, 318 pp. ISBN 0-7923-3566-X

M.S. Livšic, N. Kravitsky, A.S. Markus and V. Vinnikov: *Theory of Commuting Nonselfadjoint Operators*. 1995, 314 pp. ISBN 0-7923-3588-0

A.I. Stepanets: *Classification and Approximation of Periodic Functions*. 1995, 360 pp. ISBN 0-7923-3603-8

C.-G. Ambrozie and F.-H. Vasilescu: *Banach Space Complexes*. 1995, 205 pp. ISBN 0-7923-3630-5

E. Pap: *Null-Additive Set Functions*. 1995, 312 pp. ISBN 0-7923-3658-5

C.J. Colbourn and E.S. Mahmoodian (eds.): *Combinatorics Advances*. 1995, 338 pp. ISBN 0-7923-3574-0

V.G. Danilov, V.P. Maslov and K.A. Volosov: *Mathematical Modelling of Heat and Mass Transfer Processes*. 1995, 330 pp. ISBN 0-7923-3789-1

A. Laurinčikas: *Limit Theorems for the Riemann Zeta-Function*. 1996, 312 pp. ISBN 0-7923-3824-3

A. Kuzhel: *Characteristic Functions and Models of Nonself-Adjoint Operators*. 1996, 283 pp. ISBN 0-7923-3879-0

G.A. Leonov, I.M. Burkin and A.I. Shepeljavyi: *Frequency Methods in Oscillation Theory*. 1996, 415 pp. ISBN 0-7923-3896-0

B. Li, S. Wang, S. Yan and C.-C. Yang (eds.): *Functional Analysis in China*. 1996, 390 pp. ISBN 0-7923-3880-4

P.S. Landa: *Nonlinear Oscillations and Waves in Dynamical Systems*. 1996, 554 pp. ISBN 0-7923-3931-2

A.J. Jerri: *Linear Difference Equations with Discrete Transform Methods*. 1996, 462 pp. ISBN 0-7923-3940-1

Other *Mathematics and Its Applications* titles of interest:

I. Novikov and E. Semenov: *Haar Series and Linear Operators*. 1997, 234 pp.
ISBN 0-7923-4006-X

L. Zhizhiashvili: *Trigonometric Fourier Series and Their Conjugates*. 1996, 312 pp.
ISBN 0-7923-4088-4

R.G. Buschman: *Integral Transformation, Operational Calculus, and Generalized Functions*. 1996, 246 pp. ISBN 0-7923-4183-X

V. Lakshmikantham, S. Sivasundaram and B. Kaymakcalan: *Dynamic Systems on Measure Chains*. 1996, 296 pp. ISBN 0-7923-4116-3

D. Guo, V. Lakshmikantham and X. Liu: *Nonlinear Integral Equations in Abstract Spaces*. 1996, 350 pp. ISBN 0-7923-4144-9

Y. Roitberg: *Elliptic Boundary Value Problems in the Spaces of Distributions*. 1996, 427 pp.
ISBN 0-7923-4303-4

Y. Komatu: *Distortion Theorems in Relation to Linear Integral Operators*. 1996, 313 pp.
ISBN 0-7923-4304-2

A.G. Chentsov: *Asymptotic Attainability*. 1997, 336 pp. ISBN 0-7923-4302-6

S.T. Zavalishchin and A.N. Sesekin: *Dynamic Impulse Systems*. Theory and Applications. 1997, 268 pp. ISBN 0-7923-4394-8

U. Elias: *Oscillation Theory of Two-Term Differential Equations*. 1997, 226 pp.
ISBN 0-7923-4447-2

D. O'Regan: *Existence Theory for Nonlinear Ordinary Differential Equations*. 1997, 204 pp.
ISBN 0-7923-4511-8

Yu. Mitropolskii, G. Khoma and M. Gromyak: *Asymptotic Methods for Investigating Quasiwave Equations of Hyperbolic Type*. 1997, 418 pp. ISBN 0-7923-4529-0

R.P. Agarwal and P.J.Y. Wong: *Advanced Topics in Difference Equations*. 1997, 518 pp.
ISBN 0-7923-4521-5

N.N. Tarkhanov: *The Analysis of Solutions of Elliptic Equations*. 1997, 406 pp.
ISBN 0-7923-4531-2

B. Riečan and T. Neubrunn: *Integral, Measure, and Ordering*. 1997, 376 pp.
ISBN 0-7923-4566-5

N.L. Gol'dman: *Inverse Stefan Problems*. 1997, 258 pp. ISBN 0-7923-4588-6

S. Singh, B. Watson and P. Srivastava: *Fixed Point Theory and Best Approximation: The KKM-map Principle*. 1997, 230 pp. ISBN 0-7923-4758-7

A. Pankov: G-*Convergence and Homogenization of Nonlinear Partial Differential Operators*. 1997, 263 pp. ISBN 0-7923-4720-X

Other *Mathematics and Its Applications* titles of interest:

S. Hu and N.S. Papageorgiou: *Handbook of Multivalued Analysis*. Volume I: Theory. 1997, 980 pp. ISBN 0-7923-4682-3 (Set of 2 volumes: 0-7923-4683-1)

L.A. Sakhnovich: *Interpolation Theory and Its Applications*. 1997, 216 pp. ISBN 0-7923-4830-0

G.V. Milovanović: *Recent Progress in Inequalities*. 1998, 531 pp. ISBN 0-7923-4845-1

V.V. Filippov: *Basic Topological Structures of Ordinary Differential Equations*. 1998, 530 pp. ISBN 0-7293-4951-2

S. Gong: *Convex and Starlike Mappings in Several Complex Variables*. 1998, 208 pp. ISBN 0-7923-4964-4

A.B. Kharazishvili: *Applications of Point Set Theory in Real Analysis*. 1998, 244 pp. ISBN 0-7923-4979-2

R.P. Agarwal: *Focal Boundary Value Problems for Differential and Difference Equations*. 1998, 300 pp. ISBN 0-7923-4978-4

D. Przeworska-Rolewicz: *Logarithms and Antilogarithms*. An Algebraic Analysis Approach. 1998, 358 pp. ISBN 0-7923-4974-1

Yu. M. Berezansky and A.A. Kalyuzhnyi: *Harmonic Analysis in Hypercomplex Systems*. 1998, 493 pp. ISBN 0-7923-5029-4

V. Lakshmikantham and A.S. Vatsala: *Generalized Quasilinearization for Nonlinear Problems*. 1998, 286 pp. ISBN 0-7923-5038-3

V. Barbu: *Partial Differential Equations and Boundary Value Problems*. 1998, 292 pp. ISBN 0-7923-5056-1

J. P. Boyd: *Weakly Nonlocal Solitary Waves and Beyond-All-Orders Asymptotics*. Generalized Solitons and Hyperasymptotic Perturbation Theory. 1998, 610 pp. ISBN 0-7923-5072-3

D. O'Regan and M. Meehan: *Existence Theory for Nonlinear Integral and Integrodifferential Equations*. 1998, 228 pp. ISBN 0-7923-5089-8

A.J. Jerri: *The Gibbs Phenomenon in Fourier Analysis, Splines and Wavelet Approximations*. 1998, 364 pp. ISBN 0-7923-5109-6

C. Constantinescu, W. Filter and K. Weber, in collaboration with A. Sontag: *Advanced Integration Theory*. 1998, 872 pp. ISBN 0-7923-5234-3

V. Bykov, A. Kytmanov and M. Lazman, with M. Passare (ed.): *Elimination Methods in Polynomial Computer Algebra*. 1998, 252 pp. ISBN 0-7923-5240-8

W.-H. Steeb: *Hilbert Spaces, Wavelets, Generalised Functions and Modern Quantum Mechanics*. 1998, 234 pp. ISBN 0-7923-5231-9

Other *Mathematics and Its Applications* titles of interest:

X. Xu: *Introduction to Vertex Operator Superalgebras and Their Modules*. 1998, 356 pp.
ISBN 0-7923-5242-4

E.E. Rosinger: *Parametric Lie Group Actions on Global Generalised Solutions of Nonlinear PDFs* and a Solution to Hilbert's Fifth Problem. 1998, 234 pp. ISBN 0-7923-5232-7

MIX
Papier aus verantwortungsvollen Quellen
Paper from responsible sources
FSC® C105338

If you have any concerns about our products,
you can contact us on
ProductSafety@springernature.com

In case Publisher is established outside the EU,
the EU authorized representative is:
Springer Nature Customer Service Center GmbH
Europaplatz 3, 69115 Heidelberg, Germany

Printed by Libri Plureos GmbH
in Hamburg, Germany